0 JUN 2010
KT-405-488

CHESHIRE
COUNTY COUNCIL

About the author

Dinyar Godrej is a co-editor of the *New Internationalist* magazine.

Special thanks to:
Martina Krueger, for much help and kindness. Adam Ma'anit for good-humored support.
Troth Wells for her patience.
Bart Lienard, for putting up without complaint with endless pleas for help and much neglect.
My parents Noshirwan and Phyllis, for being lovely.

Other titles in the series

The No-Nonsense Guide to Animal Rights
The No-Nonsense Guide to Climate Change
The No-Nonsense Guide to Conflict and Peace
The No-Nonsense Guide to Fair Trade
The No-Nonsense Guide to Globalization
The No-Nonsense Guide to Human Rights
The No-Nonsense Guide to International Development
The No-Nonsense Guide to Islam
The No-Nonsense Guide to Science
The No-Nonsense Guide to Tourism
The No-Nonsense Guide to World Health
The No-Nonsense Guide to World History
The No-Nonsense Guide to World Poverty

About the New Internationalist

The *New Internationalist* is an independent not-for-profit publishing co-operative. Our mission is to report on issues of global justice. We publish informative current affairs and popular reference titles, complemented by world food, photography and gift books as well as calendars, diaries, maps and posters – all with a global justice world view.

If you like this *No-Nonsense Guide* you'll also love the *New Internationalist* magazine. Each month it takes a different subject such as *Trade Justice, Nuclear Power* or *Iraq*, exploring and explaining the issues in a concise way; the magazine is full of photos, charts and graphs as well as music, film and book reviews, country profiles, interviews and news.

To find out more about the *New Internationalist*, visit our website at **www.newint.org**

The NO-NONSENSE GUIDE to
CLIMATE
CHANGE
Dinyar Godrej

ni

The No-Nonsense Guide to Climate Change
First published in the UK by
New Internationalist™ Publications Ltd
Oxford OX4 1BW, UK
www.newint.org
New Internationalist is a registered trade mark.

First published 2001; second edition 2002, third edition 2006.
Reprinted 2007.

Cover image: Drought in Spain/NTRES REUTERS/Victor Fraile

Series editor: Troth Wells
Design by New Internationalist Publications Ltd.
Typeset by Wessex Translations Ltd.

PRINTED ON RECYCLED PAPER Printed on recycled paper by TJ International Ltd, Padstow, Cornwall, UK who hold environmental accreditation ISO 14001.

British Library Cataloguing in Publication Data.
A catalogue record for this book is available from the British Library.

Library of Congress Cataloguing-in-Publication Data.
A catalogue record for this book is available from the Library of Congress.

ISBN: 978-1904456-414

Foreword

MOST EVERYONE HAS heard about climate change, though they might not fully understand all the issues. Some are even still in denial thanks to the fossil fuel industry's efforts to sow confusion and doubt in the public debate. Many indigenous communities, however, have no doubt of its existence and very real implications for their livelihoods and the future of their way of life.

Scientists too are increasingly certain. They continue to conclude with deep urgency that climate change is creating dangerous conditions that require immediate attention.

The measurable impacts are already clear. Human survival is increasingly threatened by the bio-degenerative consequences of an industrial and technological society: depletion of aquifers and water resources; expanding deserts; decreasing forests; declining fisheries; poisoned food, water and air; and climatic extremes such as floods, hurricanes and droughts. Air pollution from burning fossil fuels causes health effects such as asthma and acute respiratory disorders. While environmental degradation in itself is by no means new, the compounding changes caused by climate change can further contribute to the dimensions of extreme poverty, hunger, human settlement, displacement, landless populations and even increases in child mortality in drought-afflicted countries such as Africa.

Communities disproportionately impacted by climate change and the continued exploration, extraction and burning of fossil fuels include small island states, whose very existence is threatened, as well as indigenous peoples, the poor and the marginalized, particularly women, children and the elderly around the world. Solutions to the climate change problem must emphasize real, verifiable reductions in fossil fuel emissions.

Foreword

Climate change has become politicized with overlays of complexities that have generally limited the search for solutions to within the ranks of politicians, lawyers, technocrats and the petroleum and energy industry. Grassroots people all over the world are starting to denounce further delays in transitioning away from a fossil fuel economy that are being caused by corporate and governmental attempts to construct a carbon market instead of legislating real change.

The No-Nonsense Guide to Climate Change provides much needed information that helps us learn about climate change and promote genuine solutions to the problem.

The connection between sustainable development and the negative effects of climate change is undeniable. The need to address climate change as a human rights issue is becoming more urgent. Climate change is a genuine threat to the health, physical and cultural survival of not just our indigenous peoples, but to all humans on this planet we call Mother Earth.

As Corbin Harney, spiritual leader of the Western Shoshone indigenous people says, 'We, the people, are going to have to put our thoughts together, our power together, to save our planet here. We've only got one water, one air, one Mother Earth.'

Tom Goldtooth
Executive Director,
Indigenous Environment Network

CONTENTS

Introduction

'WORLD GONE CRAZY – IT'S OFFICIAL!'
That could have been the newspaper headline. In July 2003 the World Meteorological Association departed from its usual detailed (and rather dull) scientific reports to issue a weather alert of sorts. It was quite a dramatic one, saying in essence that the world's weather was doing alarming things. Weather related disasters had been so remarkable that year that the WMO, which gathers input from the weather services of 185 countries, felt compelled to link them to climate change. According to the announcement: 'New record extreme [weather] events occur every year somewhere in the globe, but in recent years the number of such extreme [events] has been increasing.' The language may have been studiously detached but the message shot across the growing band of climate watchers' horizon like a distress flare.

In 2001, when the first edition of this book appeared, I re-read sections I'd written wondering if they might come across as somewhat hysterical to someone unfamiliar with the nature of global warming and climate change. But all the scientific writing I had consulted then strengthened my belief that alarm was the only natural response to the predictions. At that time I had also wondered about the wisdom of including anecdotal instances of shifts in climate and freakish weather phenomena.

Today, in 2006, the sections that seemed hysterical to me look somewhat tame compared to the increasing outspokenness of scientists on the issue and the hyperbole-laden headlines that are now common. And anecdotal instances are being reported in their thousands and inspected like the entrails of some soothsayer's unfortunate sacrifice. Today that elusive beast 'the general reader' for whom this book was written is likely to be quite clued up about what used to be called 'the climate debate'. It is, thankfully, no longer a debate. The message is now clear and direct: humanity is changing the world's climate and we need to do something about it double quick.

We have seen the images: a glacier being wrapped in Switzerland in a desperate effort to protect it from melting away; ice masses in Greenland the size of entire cities turning into gargantuan crystal sculptures as heat scoops them out. Folks washing clothes in their backyard on the low-lying island of Tuvalu, while the encroaching sea vaults onto their land. The unblinking faces from various drought-related famines in Africa. All the pieces fit – only to form a picture of a world that threatens to disintegrate.

We know this is serious. We have it from any number of authorities and we have it increasingly from our own experience. We even have it from unlikely quarters. Such as the mild-mannered diplomat and UN weapons inspector Hans Blix who asserted, 'To me the question of the environment is more ominous than that of peace and war... I'm more worried about global warming than I am of any major military conflict.'

Today it is rare to see a scientific study which revises key indicators of climate change in a reassuring direction. Instead the climate models as they increase in sophistication all make ever-worsening predictions. Some environmentalists have given up all hope of humankind's possible deliverance from the changes we have set in motion.

Yet the pace of political action on the issue has been glacial. Except perhaps glacial is the wrong word these days, seeing how fast those glaciers are going in the summer melt. And what little political action we've seen has been motivated by opportunism and greed rather than the best interests of a feverish world.

Much has been made of the destructive unilateralism of the US (the world's most polluting nation) in climate change negotiations. For too long US politicians have drummed up the threat of an economic downturn as their excuse for not addressing an issue which will make any such downturn feel like a fairground ride. For too long they have been wooed by the oil and coal lobbies into denying the overwhelming scientific consensus behind climate change. For too long they have striven to obstruct other nations' admittedly meager commitments to action.

But there is another US. It is a country where opinion poll after opinion poll has made clear that everyday people recognize the problem despite media manipulation and are willing to start making lifestyle changes. It is a country where local governments have already enacted progressive legislation to curb harmful greenhouse gas emissions despite the bloody-minded stance of the Bush administration.

In March 2006, some US media began to wake up to what the rest of the world already knew. *Time* magazine concluded: 'By any measure, Earth is at the tipping point. Suddenly and unexpectedly the crisis is upon us.' ABC News provided a week-long coverage of global warming. Could a slumbering giant be beginning to stir?

In the face of the likelihood of more and more extreme weather in our own lifetimes – not two or three generations hence – hope must be created. The changes we are witnessing today are the result of greenhouse gas emissions from over half a century ago when levels were much lower than they are today. Bearing this time-lag in mind we can expect much worse to come. So while traditional politics seems more concerned with misguided economic priorities than taking decisive action, the time is ripe for a different kind of politics. We have seen people in their thousands take to the streets to protest against the unjust indebtedness of Majority World (or 'Third World') countries and the tyrannies of the globalization of world trade. It is time to make our voices heard on the climate issue as well and let politicians and big business know that we demand real solutions while these still have a chance of succeeding.

Dinyar Godrej
Rotterdam

1 Overview of climate change

The evidence for climate change... the outlook... how 'feedbacks' could lead to runaway global warming, and the crises facing humanity.

WHEN WE ARE ill with a fever, our body temperature changes from being 'normal' at around 98.6° Fahrenheit (37° Celsius) to fever pitch at 100° F (37.8° C). A tiny increase of 1.4° F (0.8° C) switches us from feeling well to feeling sick. This sensitivity of the human body to temperature changes is one way of appreciating the phenomenon of climate change (powered as it is by global warming) which has taken up permanent residence in the world's headlines.

The world's 'normal' temperature from the frozen environs at the Poles to the noonday furnaces of some of its deserts evens out at an average of around 57.2° F (14° C). According to the late Brazilian ecologist, former Minister for the Environment José Lutzenberger, 'The range propitious to life ranges from a few degrees below zero [32° F], where life survives by resting, to about 80° C [176° F] above zero for a few organisms, some bacteria and algae that manage to live in hot springs.'[1] Lutzenberger made a comparison with possible temperatures in the universe – from approximately three degrees above -459.4° F (-273° C), also known as absolute zero, on the outer planets like Pluto and Neptune to an estimated 10,832° F (6,000° C) on the sun, to hundreds of billions of degrees within supernovas. If this range were to be plotted on a line where every degree corresponded to a fraction of an inch, or a millimeter, it would stretch to hundreds of thousands of miles or kilometers. The range within which the Earth's creatures survive would account for a mere 4 inches or 10 centimeters.

This humbling calculation is a reminder of the uniqueness of life on our planet. All life forms thrive within this small range of temperatures, in sharp contrast to the wastes of our planetary neighbors. And most life is sensitive to small changes in temperature within this relatively narrow range.

Since the 1980s temperatures have begun to accelerate. The top 20 warmest years (since records began in the latter half of the 19th century) have all been since 1981, and the top 10 since 1990. Until recently 1998 had the dubious distinction of being the hottest year on record. But a NASA team reported 2005 had overtaken it; other recording agencies put 2005 just behind 1998. The positive change in temperature over the 20th century into the present is in the region of 1.44° F (0.8° C), with 0.9° F (0.5° C) occurring since 1975. It may not seem like much, but it appears to be operating like the change that triggers a fever.

In fact the change, if one considers the enormity of the forces involved – the great landmasses, the oceans, the lower atmosphere – is not small at all. And in the last 25 years in particular, Earth has been warming at a rate that is faster than at any point in at least the last 2,000 years.

Though all the phenomena that indicate climatic change are wider than global warming, there is little doubt that this is the engine pushing these changes. The

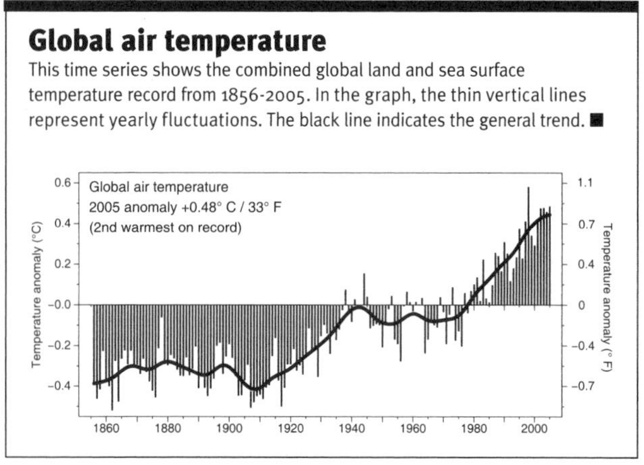

Global air temperature

This time series shows the combined global land and sea surface temperature record from 1856-2005. In the graph, the thin vertical lines represent yearly fluctuations. The black line indicates the general trend. ■

Source: Climate Research Unit, University of East Anglia, 2006.

reason for this, put very simply, is the fact that the world's weather systems are driven by the sun's energy. Each year the earth receives energy equivalent to 1,000 trillion (a trillion = a million million) barrels of oil in the form of sunlight.[2] Much of it gets reflected right back into space by the shiny surfaces of snow and ice. The rest functions as a kind of pump driving ocean currents, evaporation, snow and rain. Trees absorb some of this sunlight and release vapor by recycling the water they have taken up from the soil, further fueling the constant movement of life-giving water on our planet. Brazilian physicist Eneas Salati estimates that the energy flow each day over the Amazon Basin alone is comparable to between five and six million atom bombs.[3] Despite these huge, almost unfathomable quantities, the world's weather systems are quite finely tuned entities – knock them and they could tip out of balance. Add more energy into the equation in the form of global warming and there are bound to be corresponding accelerations and imbalances.

Detecting warming

A warming of 1.44° F (0.8° C) in global terms has all the regional variation and complexity that we can expect from, say, the weather in different parts of the world. For one, it has been uneven, with the greatest variations being over land in the middle-to-high latitudes of the northern hemisphere, although localized areas such as the southern Mississippi Valley have actually cooled. Worldwide, temperatures have crept up most during nighttime and have made the strongest rises during the northern winter and spring months. US winters are breaking records for being warmer, a trend which seems to be emerging in Australia as well. The poles have witnessed freak changes, with recording stations registering rises of up to 9° F (5° C).

Whilst temperature records go back only for 150 years, scientists have been excavating the earth's temperature going much further back using 'proxy methods', taking measurements from tree rings, deep ice cores, sea sediments and ground bores. What these prehistoric records reveal is staggering – at no point in the last 2,000 years has the earth's temperature changed as rapidly as in the 20th century.

Tracing the changes

Today the idea that human activities could be fueling this change in temperatures and with it changes in the world's weather has wide currency. But to arrive at that conclusion one needs to account for other perfectly natural possibilities that can cause the climate to vary.

Over the lifetime of the Earth, there can be little doubt that variations in the amount of solar energy given out by the sun have changed climatic systems considerably. If tomorrow the sun's output of energy increased even fractionally, it would result in quite radical climatic changes on Earth. But for the warming observed through the past century, perhaps the most persuasive natural argument was that put forward by Knud Lassen of the Danish Meteorological Institute. He proposed that the 11-year cycle of sunspot activity on the face of the sun seemed to synchronize with trends in global temperatures, a theory which was a particular favorite amongst those who wished to believe that nothing could be done about global warming. But in 2000 Lassen admitted weaknesses in his hypothesis, telling a meeting of the European Geophysical Society that sunspots and solar cycles couldn't explain the dramatic surge in temperatures since 1980. His colleague Peter Thejll said, 'The curves diverge after 1980, and it is a startling deviation. Something else is acting on climate. It has the fingerprints of the greenhouse effect.'

The United Nations' Intergovernmental Panel on Climate Change (IPCC), the world's leading body on the subject, has also studied solar activity closely. It found that despite the sun's increased radiance during the first half of the 20th century, which led to a small increase in the amount of solar energy hitting the earth's surface, this alone could not account for the rising temperatures. In fact the cooling impact of volcanic activity (see below) over the last century would have more than wiped out the effects of this increase in solar energy. If anything the closing decades should have seen a decrease in temperatures.

Another theory, posited in the 1920s by Serbian meteorologist Milutin Milankovitch, claimed that changes in the tilt and orbit of the earth over periods spanning millennia could cause climatic changes because they would affect the way the sun's energy was distributed on different parts of the planet. The 100,000-year shift in the earth's orbit around the sun could, according to Milankovitch, be the underlying cause of our planet's cycles of ice ages. However, Milankovitch's cycles describe changes occurring over a time frame of hundreds of thousands of years and are not enough in themselves to explain the making of ice ages. One has to invoke other factors such as a decrease in atmospheric levels of the greenhouse gas carbon dioxide (see below).

Volcanic explosions also have the capacity to shake the system – cooling the earth rather than warming it. They throw up vast clouds of dust and sulfur dioxide into the lower atmosphere. The dust eventually settles or gets rained out, but the sulfur dioxide spreads a cloak of pollution, which reduces the amount of the sun's energy that gets through. Volcanic eruptions can cool the earth's temperatures by about 0.4-0.5° F (0.2-0.3° C). But such effects in the 20th century have rarely lasted for more than a few years, so they cannot be held responsible for changes over the longer term.

The overheated engine of human progress

All of which points to another likely culprit – human influence. The havoc that humans are playing with the weather ironically has the stuff of all life on Earth as its basis. This is the element carbon, the essential building block of anything that breathes or grows, whether animal, vegetable, or in-between. Earth is packed with it in the form of their remains. But it also envelops the planet in the form of the gas carbon dioxide, invisible but with one remarkable quality that it shares with a few other gases. It allows the sun's energy to reach the earth's surface relatively unhindered, but traps a certain amount of energy re-emitted by the warmed earth (which is of a longer wavelength), before it can escape to space. By bringing about this delay it performs the ultimate life-giving function, keeping the world's temperature on an even keel. The dance of carbon in the air is a remarkable one – we exhale it with every breath, it is liberated through volcanic explosions, fires and the decomposition of dead things. And, on the other hand, plants both on land and in the sea draw it back down in order to make their food. They later release some at night when they respire and the rest is liberated when the plants die or are burnt or get eaten by animals which then produce methane. But for the duration of their life cycles – which for some trees

The greenhouse effect

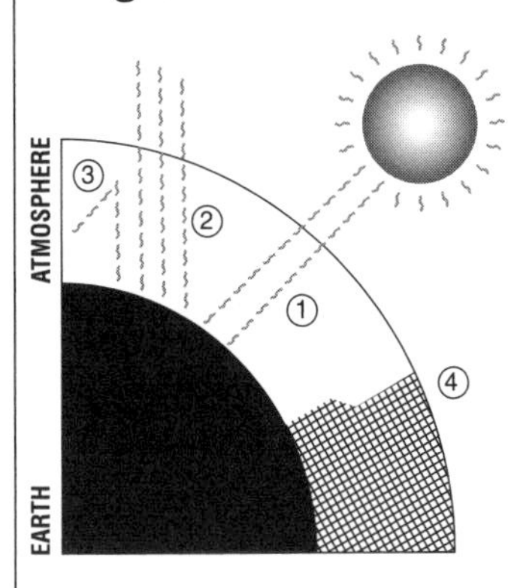

A layer of gases in the atmosphere acts like an insulating blanket trapping solar energy that would otherwise escape into space. Without these 'greenhouse gases' the earth would be frozen, barren and lifeless.

How it works

1 Solar energy enters the atmosphere at a wavelength short enough to allow it pass unaffected by greenhouse gases.

2 The sun's rays are absorbed by the earth then reflected back at longer heat wavelengths.

3 Greenhouse gases absorb some of this heat, trapping it within the lower atmosphere.

4 When greenhouse gas concentrations increase, more heat is captured causing temperatures in the lower atmosphere and surface to rise. This affects both weather and climate. ■

can be hundreds of years – they store carbon on earth. The waters of the oceans – which are the only true sink of carbon dioxide and the only net producers of oxygen – are constantly soaking it up. In fact the carbon stores of the oceans are around 50 times greater than the entire amount of carbon in the atmosphere. It's a system of endless recycling – or at least it should be.

Sadly a vital component of human advancement since industrialization began, aided by greed, is throwing the system off-balance. By burning fossil fuels (coal, oil and natural gas) to supply our power and manufacturing needs and to drive our cars, we're adding around seven billion tons of carbon (equivalent to 26 billion tons of carbon dioxide) to the atmosphere each year. This is a fraction of the amount already up there, but extra nonetheless, with about half not getting reabsorbed, thus building up year on year. Its levels in the atmosphere could double pre-industrial levels by 2080 (though many commentators fear we may not have to wait that long, some putting the year by which this happens as early as 2030). Alongside this the chopping down of forests continues unabated, diminishing some of Earth's ability to soak up carbon, to say nothing of the methane released from exposed soils and decomposing roots.

Climate modeling

The question that has vexed many minds is of the degree of confidence with which we can attribute climate change to such human activities. After all, whatever the human contribution may be it is bound to be 'superimposed on, and to some extent masked by, natural climate fluctuations'.[4] How then to distinguish the human signal from the background noise of natural fluctuations? In order to do this, a change that diverges from the normal range must first be detected and then plausible causes found that can explain that change.

The relationship between carbon dioxide and the warming of our planet's atmosphere is nothing new. Swedish scientist Svante Arrhenius had suggested back in 1896 that increased levels of the gas in the atmosphere would lead to warming. But his theories only got dusted down again when attention started turning towards climate change in the 1980s. At that time climate modeling was being more fully developed with the aid of computers. Building such models is an extremely complex business and yet comparatively lacking in sophistication when compared with the complexity of all the factors that actually contribute to the climate. Whereas short-term weather forecasts use satellite and ground data of observable weather phenomena to give a reasonably accurate short-term prediction, climate modeling looks at the longer term-frame of seasons and years and gives the broad outlines for them.

In fact the IPCC – which has input from over 2,000 of the world's leading scientists working in various fields – has been constantly refining models to improve the accuracy of the forecast. The British Met Office's modelers work with some 300 terabytes (million million bytes) of data, a number which is expected to double each year. The quantities of variables fed into such models strain the abilities of the supercomputers used for this work. And naturally, as the amount of anthropogenic (human-caused) emissions of greenhouse gases in the future could vary depending on whether pollution controls are imposed or not, climate models also have to take into account different emissions scenarios and make predictions accordingly.

Rather than basing the analysis on overall warming, the models take into account patterns of change over different geographical regions and over the seasonal variations. This is because the causes of climate change show different patterns of climate response, which the modelers try to match with changes that have already been observed. Confidence in the models was significantly boosted when they were run backwards over the past century and their predictions gave the correct general outline of climatic changes for that period. A climate model also correctly predicted the damping of temperatures and period of recovery after the eruption of Mount Pinatubo in the Philippines in 1991.

The IPCC's findings make for disturbing reading, none more so than the temperature increases that have been forecast, with an increase of a massive 10.6° F (5.9° C) on the cards by 2100 if nothing is done about greenhouse gas emissions. A sobering light is shed on this estimate by the fact that a slightly smaller rise in temperature accounts for the difference between the last Ice Age when much of the northern hemisphere was buried thousands of feet deep in ice and the situation today. Another climate prediction project co-ordinated from Oxford University stretched the upper limit to a mind-boggling 19.8° F (11° C) – just a few degrees short of the planet's current average temperature.

As carbon dioxide (CO_2) is the most abundant greenhouse gas, it is the biggest player in the climate change stakes, contributing over 60 per cent to the global greenhouse effect. Much research has gone into establishing that the increase in CO_2 is really due to human activity. Fortunately there is a fail-safe way of establishing this. The nuclei of carbon atoms in the gas in emissions from natural and human processes are different. Naturally released carbon dioxide's carbon atoms have a measurable level of radioactivity. But the nuclei of carbon from fossil fuel sources have lost this radioactivity after being buried deep in the earth for millions of years. Tree rings have provided the evidence that concentrations of radioactive carbon-14 are getting diluted, which means there is more carbon dioxide from the burning of fossil fuels around.

Scientists have also been measuring the amounts of the gas in the atmosphere since the 1950s and they have observed an increase in its levels with each successive year. Ice cores from Greenland and the Antarctic ice caps provide a record of the atmosphere going back 200,000 years in the tiny bubbles of air trapped in the ice when it first fell to the ground as snow. These same air bubbles also contain a record of temperatures long before direct measurements

The major greenhouse gases

Without its protective atmosphere the earth would be 71.2° F (33° C) colder – frozen and lifeless. But over 99 per cent of the atmosphere – composed of nitrogen and oxygen – does not retain much of the sun's heat. The gases that cause the natural greenhouse effect, trapping some of the sun's heat and making life possible form just a tenth of one per cent of the atmosphere's total volume.
Human activities have sent levels of greenhouse gases soaring. Fledgling attempts to regulate major climate-damaging emissions have culminated in the UN's Kyoto Protocol to the Framework Convention on Climate Change, which was agreed in 1997 and came into force in 2005. The Protocol covers six major gases.

- **Carbon dioxide (CO_2)**

Current levels are higher than in the last million years. While nature produces about 30 times more than humans as part of a finely tuned cycle, we still spew over 22 billion tonnes* of the gas each year mainly through the burning of fossil fuels. Land-use changes such as deforestation and clearing of land for logging, ranching and agriculture account for 15 to 20 per cent of human emissions. These additional emissions accumulate and could hang around for hundreds to even thousands of years warming the planet.

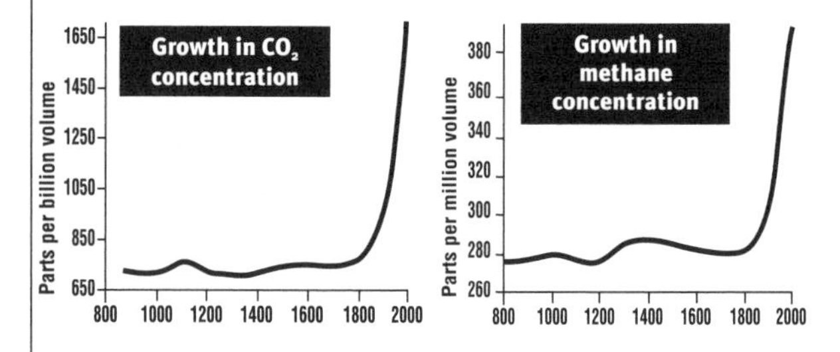

- **Methane (CH_4)**

62 times more powerful than carbon dioxide in terms of its global warming potential over 20 years; 23 times over a longer time-span of 100 years. Methane levels have risen a full 150 per cent above pre-industrial levels. This gas is created through deforestation, decomposition of waste, and rice and cattle production and stays around for approximately 12 years. If the permafrost (the part of the earth's surface, primarily in arctic regions, that remains permanently frozen) starts melting, 400 billion tonnes* of methane could be released. There is scientific debate about how methane levels in the atmosphere are measured with some scientists believing that its role has been underestimated.

• **Nitrous oxide (N_2O)**

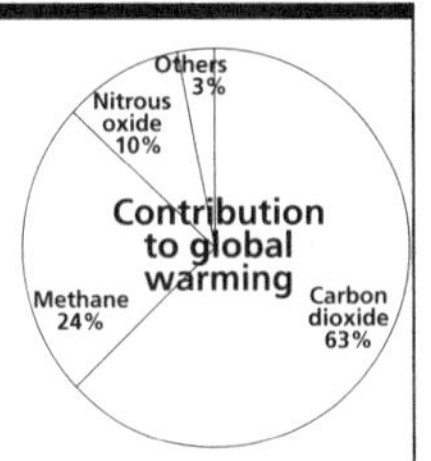

Heavy use of nitrogen fertilizers in industrial agriculture, production of synthetic materials such as nylon, and the burning of fossil fuels, has increased levels of this gas by 16 per cent of pre-industrial levels. Nitrous oxide is 296 times more powerful than carbon dioxide over a 100 year period and stays around for 120-150 years.

F-gases

Fluorinated gases, or 'F-gases', are potent greenhouse gases primarily used in industrial applications. Even though they account for a small proportion of overall emissions, there is concern that, if left unchecked, F-gases could equal at least 15 per cent of total greenhouse gas emissions in 2040 and 40 per cent by 2100. Three families of F-gases are regulated under the Kyoto Protocol. These are HFCs, PFCs, and SF_6.

• **Hydrofluorocarbons (HFCs)**

HFCs are primarily replacements for ozone-depleting substances that are being phased out, like chlorofluorocarbons (CFCs) and hydrochlorofluorocarbons (HCFCs). Recently there has been sharp growth in the use of HFCs, which has raised concerns since they are among the most potent greenhouse gases with a global warming potential significantly higher than that of carbon dioxide (HFC-23, for example, is 12,000 times more potent as a greenhouse gas than CO_2 and can remain in the atmosphere for 260 years). HFC is used primarily in refrigerants, aerosols, and industrial applications such as in the manufacture of certain solvents and materials such as the plastic PTFE (more commonly known as Teflon).

• **Perfluorocarbons (PFCs)**

Similarly to hydrofluorocarbons, perfluorocarbons are sometimes used as a replacement for ozone-depleting CFCs, particular in refrigerants. A major source of PFCs is from the aluminum industry as a byproduct of the smelting process. Two major PFCs (CF_4 and C_2F_6) have a global warming potential of 6,500 and 9,200 times respectively to that of carbon dioxide and can last in the atmosphere for over 10,000 years.

• **Sulfur hexafluoride (SF_6)**

Of all the fluorinated gases, sulfur hexafluoride has the highest global warming potential (22,200 times that of carbon dioxide) and can last in the atmosphere for 3,200 years. It is used in large-scale electrical equipment (such as those used in nuclear power installations) as an insulator. It is also used in the 'Air Sole' range of Nike sneakers, and in sound-insulating glass. ■

**1 US ton = 2,000 lbs. 1 metric tonne = 2,240 lbs/1,000 kg.*

Sources: Hadley Centre, *Climate Change and the Greenhouse Effect*, December 2005; David Archer, 'Fate of Fossil Fuel CO_2 in Geologic Time', in *Journal of Geophysical Research*, Volume 110, 2005; IPCC, 2001, www.grida.no/climate/ipcc_tar/wg1/248.htm ; Dave Mussell, Juleta Severson-Baker, Tracey Diggins, *Climate Change: Awareness and Action* (Pembina Institute for Appropriate Development, Ottawa, 1999).

began. In pre-industrial times carbon dioxide levels were around 280 parts per million volume (ppmv), in 2006 they had crossed 380. According to the British Government's chief scientific advisor David King: 'That's higher than we've been for over a million years, possibly 30 million years.'[5] If nothing is done to curb carbon dioxide emissions they could double before the century is out.

Evidence has also come from the measurements of CO_2 at different locations, showing slightly higher readings for the northern hemisphere. As the majority of fossil-fuel burning activities are taking place in the North and atmospheric circulation patterns delay the arrival of northern emissions to the southern hemisphere by about a year, the pattern is entirely consistent.

According to James Zachos, Professor of Earth Sciences at the University of California, all greenhouse gases are being released into the atmosphere 30 times faster than when the earth experienced an episode of natural global warming after the dying out of the dinosaurs 55 million years ago. 'The emissions that caused this past episode of global warming probably lasted 10,000 years,' he said. 'By burning fossil fuels, we are likely to emit the same amount over the next three centuries.'[6]

All change

So what's in store? The general predictions of the IPCC scientists include a greater degree of warming over landmasses where people live than over seas, as the darker, rougher surfaces of land soak up solar energy better. The Arctic will see the greatest amount of warming in its winter temperatures. Night-time warming will be greater than that for daytime. For the mid-latitudes (most of North America, Europe, parts of South America) the number of hot days in summer will increase while exceptionally cold days will decrease in number. To quantify this, we can look at the European heatwave of 2003 during which 35,000 people died. According to the Hadley Center for Climate Prediction and Research: 'In the absence of any human modification of climate, temperatures such as those seen in Europe in 2003 are estimated to be a 1-in-1,000 year event. Despite this, by the 2040s, a 2003-type summer is predicted to be about average, and by the 2060s it would typically be the coldest summer of the decade.'[7]

The most worrying predictions though concern the unpredictable – there will be an increase in extreme weather events such as freak floods and prolonged droughts and they will last longer. Not all the changes would have to do with warmer weather as such; the increased energy flows could drive more intense blizzards and snowstorms as well. With temperatures rising there will be increased evaporation both from land and sea. The latter could translate into increased rainfall. But this will not be an orderly change, with the rainfall compensating for the drying out of the land. Instead there will be all sorts of local imbalances, some areas turning to desert after their soils have been baked dry whilst others see topsoil washed away by floods. The World Water Council announced in 2003 that 'major flood disasters' had risen nearly fourfold from the 1970s to the 1990s. Storms are already becoming more frequent and intense. Coastal regions could be awash with rain while great continental interiors dry up.

Many of the predictions assume gradual global climate change, but the distant past offers evidence for abrupt switches in climate. And the fear is that climate systems could undergo just such a dramatic flip, from which recovery would be a matter of centuries rather than decades (see box: Northern Europe and the big chill).

While one cannot link any particular extreme weather event with certainty to global warming, the frequency of record-breaking events each year since the 1990s has made commentators invoke it with frightening regularity.

Feedbacks that could be paybacks

Particularly alarming are the possibilities of indirect effects of this warming that could further accelerate climatic changes. These are perhaps misleadingly referred to as 'positive feedbacks' and they remain difficult to factor in to climate models.

With the warming of the oceans and the surface air above them, evaporation increases, adding more water vapor into the air. Water vapor is in fact the most abundant natural greenhouse gas and a potent one. Any increase of water vapor, caused indirectly by warming due to increases in other greenhouse gas concentrations, would further trap heat.

Northern Europe and the big chill

On the Lofoten Islands off the coast of Norway, mists rise above the sea in winter, bathing the scene in an air of dreaminess. Although these islands are within the Arctic Circle they enjoy a much more equable climate than might be expected. The mist is the reason. It forms because the sea is several degrees warmer than the air above it, causing the moisture in the air to condense. Lofoten owes its inhabitability to the warming presence of the Gulf Stream, an ocean current that keeps temperatures in large parts of Northern Europe far milder than they would otherwise be.

But global warming could have a nasty surprise for this region – far from turning it into some Mediterranean-type haven there is the possibility that the climate here could resemble that of Siberia in the span of a century or two. The reason for this would be the moving southwards of the Gulf Stream due to vast quantities of freshwater melting from the warming North Pole. More dramatically the Gulf Stream could stall and even switch off, causing winter temperatures to plummet by 18° F (10° C) or more. This could happen because the pump of the Gulf Stream works in the freezing waters of the Greenland Sea. As the seawater begins to freeze here, the remaining water becomes denser, with a much higher salt concentration. This colder, saltier water sinks slowly to the ocean floor and begins its tortuous journey to the South Pole, thus pulling warm water from the tropics up towards the north. The Gulf Stream is part of a gigantic conveyor belt of ocean currents that straddles the globe from Pole to Pole. But if increasing quantities of melt water at the North Pole kept the sea water diluted then it could gradually stop sinking, slowing the pump down, possibly leading to a point where the pump could switch off.

Current predictions for the weakening of the Gulf Stream show that global warming would override any cooling. But if the Stream were to stop the entire region would get long, bitterly cold winters and the ripple effects could be wide-ranging – even disrupting the Indian monsoon continents away. ■

Sources: Hadley Center, *Climate Change and the Greenhouse Effect*, December 2005; Mark Rowe, 'Global Warming to leave UK out in the cold', *The Independent on Sunday*, 8 October, 2000.

Once positive feedbacks are triggered, they could go on to trigger others, leading to runaway warming. Here is a hypothetical but not entirely improbable doomsday scenario. As greenhouse gases build up in the atmosphere, temperatures rise. Soils start to dry out and release carbon (this is predicted after 2050 if nothing happens to change current trends). Forests begin to die back or burn. The temperature jumps by an average of 14.4° F (8° C) over land. Areas under ice melt to expose the Earth below, which begins to soak up the sun's heat instead of reflecting it. Long-frozen tundra-vegetation begins to thaw and decompose releasing much of its methane (a potent greenhouse gas) store of 400 billion tonnes. The seas, swollen by rising temperatures and melting polar ice, swallow densely populated coastal regions. With warming they also begin to lose their ability to absorb carbon and could start releasing the gas already dissolved in them – estimated at 50 times the amount contained in the atmosphere... and so on. As vicious circles go this one is hard to beat.

As the 20th century closed, warming accelerated, with average global temperatures jumping by half a degree Celsius (0.9° F) in the last 25 years. This would be the equivalent of 2° C or 3.6° F per century. However the amount of change to which ecosystems can adapt is estimated at a maximum of 1° C (1.8° F) over a century. And that is if no further changes are expected. There could be worse to come as today's effects are believed to be mainly the work of carbon dioxide emitted half a century ago. The much higher levels of emissions today are damage we are storing up for the future. Also to be factored into the equation are the effects of sulfate aerosols, by-products of industrial pollution which have masked the greenhouse effect by their cooling properties – a phenomenon known as global dimming. They have a short atmospheric lifetime and as cleaner production processes become more desirable in our increasingly polluted world, their role could well decline, revealing the

Extreme weather, 2005

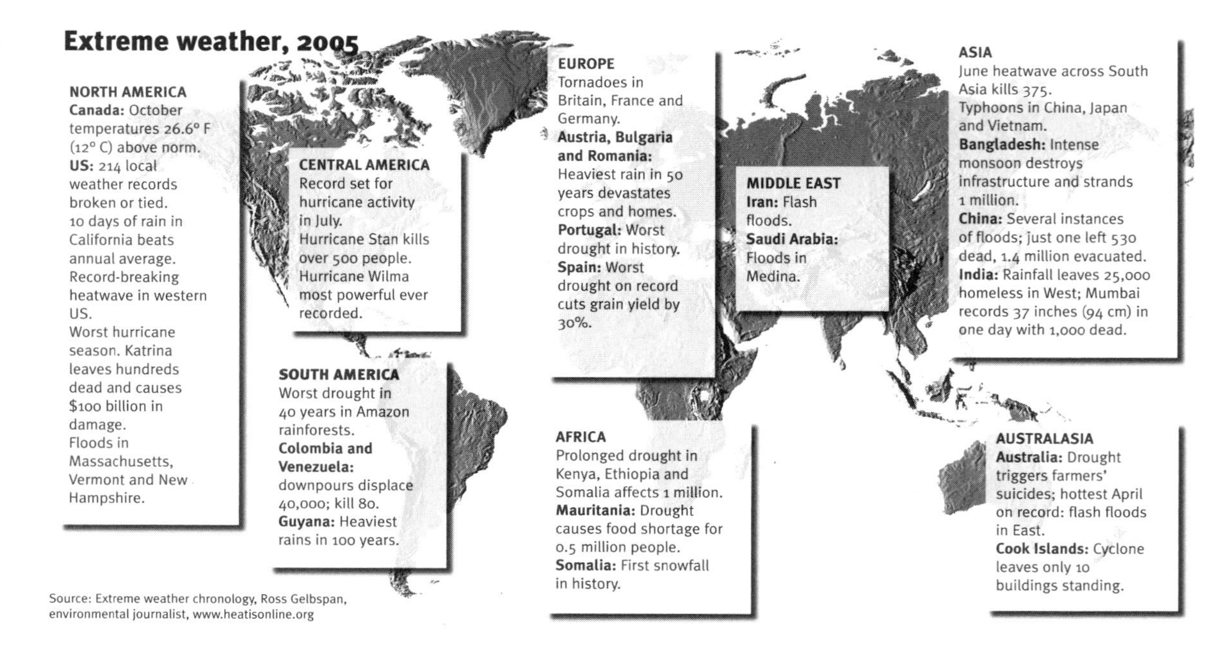

Source: Extreme weather chronology, Ross Gelbspan, environmental journalist, www.heatisonline.org

true extent of warming. However, this is still being hotly debated in the scientific community.

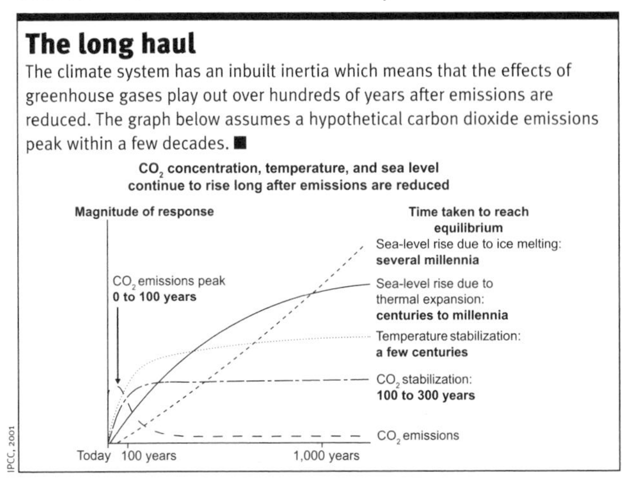

The long haul

The climate system has an inbuilt inertia which means that the effects of greenhouse gases play out over hundreds of years after emissions are reduced. The graph below assumes a hypothetical carbon dioxide emissions peak within a few decades. ■

Carbon store

The attack on the world's forests has left us with just a third of the land functioning as an intact forest ecosystem than was once under forest cover. Trees store carbon and draw it down to produce food, but dead, burnt or cleared forests turn from being carbon sinks to carbon sources. Two and a half acres (one hectare) of tropical rainforest contains between 100 and 250 tonnes of carbon in the form of organic matter (a figure which is much higher if one includes the carbon stored in the soil), three-quarters of which could be released by burning or decomposition.[3]

With such forests disappearing at the rate of one per cent a year throughout the 1990s and countries often being forced to exploit the economic potential of their timber to service their international debts the outlook is less than bright. In 2005 the Amazon rainforest suffered a freakishly severe drought and huge swathes of the forest went up in flames. The Amazon rainforest has often been called Earth's lungs because of its circulation of oxygen and its function as a pump for energy released through transpiration which leads to rain-bearing clouds that benefit areas as far-flung as Northern Europe and Scandinavia. The feedback mechanism from rainforest destruction would not only directly impact on warming but could also result in a decline in rainfall in these areas.

Another potential feedback is the release of methane from methane hydrates. These resemble ice in that they are solid, but they are actually an unstable mixture of water and methane forming at low temperatures under considerable water pressure in the oceans. An essential factor for their formation is the presence of a thick enough layer of sediments to generate the methane in the first place. Released from the ocean's pressure they sizzle and disappear into the air in a matter of seconds. The US Geological Survey estimates that the total carbon lurking in methane hydrate deposits is around 10,000 billion tonnes, most of it too far below the ocean's surface to be released. But in the Arctic with its colder water temperatures, less pressure is required for the hydrates to form and as a result they occur in much shallower waters and could conceivably be destabilized if the water warms enough.

As author, earth scientist and solar energy advocate Jeremy Leggett puts it, 'The question is, how much hydrate lies around the Arctic? We do not know for sure, but it must be measured in many tens of if not hundreds of billions of tonnes. And since there is a mere five billion tonnes of carbon in today's stock of atmospheric methane, only a little methane hydrate would need to be melted to boost the greenhouse effect significantly.'[8]

The potential of such events starting a chain of runaway feedbacks, however distant it may appear, needs to be acknowledged. And as long as that potential remains we need to act, because once the juggernaut starts rolling it is too late to stop it. This is known as the 'precautionary principle' in the climate debate and the time to apply it is now. For with each bit of corroborative evidence to support global warming, the outline becomes clearer. Waiting for the picture to be complete would be waiting for catastrophe. This section has dealt with only a few of the possible feedbacks. But the warning signs are there for anybody to see.

The big melt

In recent months the news media has been awash with headlines about the astonishing rapidity with which polar ice has been melting.

As this big melt becomes a reality, thoughts turn to the fears of rising sea levels. In 2001 the IPCC predicted a sea level rise of up to 3 feet (90 centimeters) by 2100 in its worst case scenario. A rise of less than half that would mean that annual floods could threaten an estimated 94 million people – up from the 13 million at present. The coastal regions of southern and Southeast Asia would be the worst affected, as storm surges could push sea water deep inland.[9]

For about 40 low-lying island nations worldwide the combination of fiercer storms and sea-level rise could spell complete disaster. Already the United Nations Environment Program (UNEP) has recommended the evacuation of Tarawa atoll, part of the Pacific island nation of Kiribati. Some small islands fringing Kiribati have disappeared under the waters. Roads have had to be moved inland on the main island as the ocean gnaws into the shore. In June 2000 New Zealand/Aotearoa made a promise of sanctuary to inhabitants of Tuvalu if their coral-atoll sank under the sea.

These are the canaries in global warming's coalmine. The sea level rise that figures in the predictions is due by and large to the thermal expansion of the warmer sea-waters. Now account also needs to be taken of the vast volumes of water that could be released from the world's large ice masses.

The Antarctic Peninsula has reported a sustained warming as high as 4.5° F (2.5° C). In places scientists have observed that rocks that have been covered by ice for millennia have begun to poke through. In the mid-1990s a roughly 5,000 square-mile (8,000 square kilometers) ice-shelf, Larsen A, broke away into the sea, looking – as an observing scientist put it – 'like bits of polystyrene foam smashed by a child.' Larsen B, about a fourth in size, followed in 2002.

Up north, the Arctic ice sheet is undergoing unprecedented summer melts, which aren't fully recovered in the winter. An earlier prediction that the sheet would vanish completely in summers by 2070 now looks optimistic. This would be a catastrophe for the wildlife the sea ice supports but it wouldn't make sea levels rise as the ice lies upon water to begin with. But the loss of land-based ice is another story. The Arctic's Greenland ice sheet has had more than 3 feet (1 meter) shaved off it each year since 1993 along its southern and eastern edge. Now vast segments are beginning to crumble as melting water sinks below the ice and acts as a lubricant to slide the mass above into the sea. Earlier predictions had estimated that with a 5.4° F (3° C) rise in summer temperatures, Greenland's ice would take 1,000 years to melt. It contains enough water to raise sea levels by a catastrophic 23 feet (7 meters).[7] So the speed up in the ice plunging into the sea is ringing alarm bells.

In Antarctica, three great ice sheets have gone completely, but whether land ice is also melting as rapidly is a matter of dispute. Some studies have suggested that the smaller of the continent's two land ice sheets (with a mass the size of Mexico) is melting faster than normal. If the West Antarctic ice sheet were to collapse, the seas would surge by 18 feet or 6 meters.[7] To put this into perspective, it is estimated that a 3-foot or 1-meter rise in sea level could flood many of the world's major coastal cities, such as New York, London and Bangkok and swallow up three per cent of total land area. Significantly, 30 per cent of the world's croplands could be lost.[10]

In the Netherlands, the famous sea defenses would no longer be able to protect large areas that are under sea level. In April 2000 the inhabitants of the coastal village of Bergen aan Zee were told that the authorities could no longer guarantee their safety due to rises in sea levels.

And it's not just sea ice that is melting – all over the world glaciers are in retreat as well, because the summer melt is more than can be replenished. The Quelccaya glacier in Peru is retreating 10 times faster than a decade ago, threatening water supplies for Lima's 10 million people.

Himalayan glaciers are, according to a UN study, 'receding faster than in any part of the world'. As environmental journalist Fred Pearce put it, 'Their eventual disappearance is a potential catastrophe for the hundreds of millions of people in southern Asia, who depend on the summer melt even more than the monsoon rains to irrigate their crops and provide drinking water.'[11]

For the 6,000 people whose lives are at risk from the brimming Tsho Rolpa glacial lake in Nepal, the situation is more urgent. Dozens of such lakes have formed high in the Himalayas in recent history, with only the debris left behind by retreating glaciers damming them in. About every three years, one bursts sending a wall of water rushing down the valleys. For the villagers in striking distance of Tsho Rolpa, every summer since 1994 when the alarm was first sounded has been a tense one. Work is underway to try and siphon out some of the water, but until the situation becomes safer a warning siren triggered by sensors is the only defense the villagers have. It would give them a few precious minutes' notice that the deluge was on its way.

Looking for refuge

In Bangladesh there is a widespread acceptance that seasonal floods are worsening and people have built raised fields with freshwater ponds stocked with fish to help them sit out floods. But such preparedness begs a vital question: with Bangladesh's fertile coastal plains at risk of sinking under the sea by the end of the century, who and what will feed the dispossessed people? Already the number of environmental refugees worldwide has been estimated at 25 million, more than the total of all other refugees. Fleeing into adjoining lands they are largely invisible to the rich West.

The UK charity Christian Aid calls these events unnatural disasters, because they believe there is nothing natural about them, with climate change at their root. These are staggering events like the October 1999 super cyclone that hit Orissa in India killing 30,000 and displacing 10 million others, or the failure of the rains for the fourth consecutive year in the Horn of Africa in the summer of 2000 which put 16 million people at risk of starvation. Drought has returned regularly to East and Southern Africa since.

'Country after country is being decimated by these so-called natural disasters,' said Malcolm Rodgers, Christian Aid's senior policy officer. 'The terrible irony is that the poorest countries are suffering, and we believe that this is because of pollution by the wealthiest.' People in industrialized countries generate over 62 times more carbon dioxide pollution per person than people in the least industrialized countries.

The International Federation of Red Cross and Red Crescent Societies has also been pointing out in recent editions of its annual *World Disasters Report* how the fallout from climate change has been disproportionate in the developing Majority World where people are often living on marginal lands and struggling against great economic odds to eke out a living. The 1999 edition stated that 96 per cent of all deaths from natural disasters happen in developing countries. In the period 1987 to 1996, 44 per cent of all recorded floods were in Asia, yet they caused 93 per cent of all deaths.

When 1997-98's supercharged El Niño phenomenon carved out a path of destruction across the globe, it claimed 21,000 lives as a result of floods, forest fires, droughts and disease. Most of these people lived in developing countries. While weather instability can hit any part of the world, poorer countries often have few defenses and count their losses in lives rather than insurance.

Sometimes protective measures can be far beyond the means of the poor. The cost of protecting a wealthy nation such as the Netherlands from a 20-inch (50 centimeters) rise in sea levels has been calculated at an astronomical $3.5 trillion. For the Maldives, already facing danger from the sea, the costs are real. The cost of protecting their shorelines currently runs at over $4,000 per foot ($13,000 per meter) of coast.[12]

The series of weather disturbances known as El Niño occur every three to seven years and have always trailed disaster in their wake. But commentators argue that with climate change manifesting itself in more severe weather events, it is also blowing up the manifestations of the El Niño effects. There is also the possibility that an overheated world climate system could result in El Niños returning within much shorter time scales.

The destruction that El Niño brings can be further amplified by ecological damage and poverty, as was the case when Hurricane Mitch hit Honduras and Nicaragua at the tail end of October 1998, killing some 11,000 people. Mitch had been downgraded from a hurricane to a tropical storm when it made landfall in Honduras. But with winds forced to rise by mountain ranges, it dropped a year's rainfall in just two days. In its path were flimsy homes crowded onto marginal lands that had

El Niño

El Niño is a series of weather disturbances that typically cause storms and flooding over the Pacific coast of the Americas, while Southeast Asia and the western Pacific region suffer drought. The disturbances get their name from the Spanish for 'the Christ child', which is how Peruvian fishers named the phenomenon as it usually peaks around Christmas. The phenomenon is caused by a change in sea-surface temperatures and of atmospheric pressure – during the 1997-1998 El Niño, sea temperatures were up to 9° F (5° C) higher than normal – in the tropical Pacific Ocean region. The onset of an El Niño is usually first observed when warmer waters off the coast of Peru lead to a sharp decrease in the anchovy catch. ■

been stripped of any vegetation that could have held soil together. They were soon buried under the million or so landslides that occurred. For heavily indebted Honduras, the damage in money terms was equal to 60 per cent of its annual gross domestic product. 'We lost in 72 hours what we have taken more than 50 years to build,' said the Honduran President. The same year China suffered one of the worst floods in its history when the Yangtze River basin became inundated. The damage was estimated at $30 billion.

When insurance giants tremble

Proof of the increasing violence of the world's weather is also coming from the growing jitteriness of the world's insurance giants who are outspoken on the issue of climate change.

Assessments by Munich Re, one of the world's largest insurance companies, show natural disasters doubling in frequency every decade in recent years. Of course insurance remains largely a luxury of the rich nations, but with billion-dollar catastrophes turning up with alarming regularity in the 1990s, there is concern that future extreme weather events could bankrupt the industry and destabilize world markets. In the US, catastrophe-related losses had grown from about $100 million a year in the 1950s to $6 billion per year in the 1990s. In 1992 Hurricane Andrew drove 11 insurance companies to bankruptcy despite missing vulnerable Miami. One night's fury left a tenth of the insurance industry's global reserves exhausted and resulted in new laws governing the sale of insurance in Florida. There were consequences in the Caribbean too with companies withdrawing or refusing cover against hurricanes. The first decade of the 21st century has cranked up the damage. In 2004, weather related losses totaled $145 billion, with claims reaching $45 billion. But in 2005, the year of Hurricane Katrina among others, financial losses had vaulted to $200 billion with claims running at over $70 billion.

Hurricane force

Weather instability can hit any part of the world – but already vulnerable areas are in the greatest danger. Ninety-six per cent of all deaths from natural disasters happen in developing countries. People living in insecure housing on marginal land are most at risk. ■

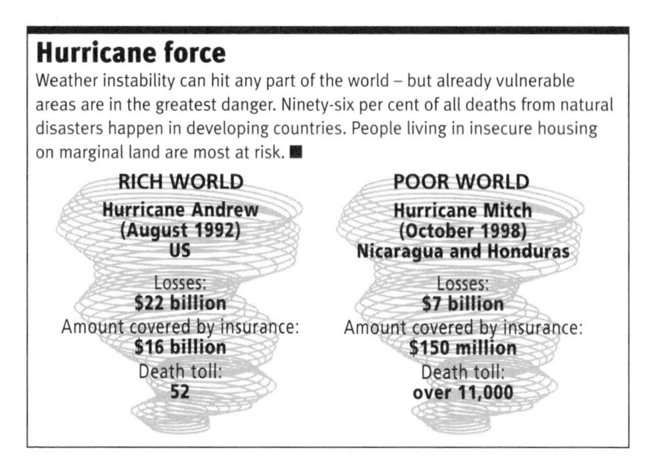

Ultimately no amount of insurance could shore up those trapped in the midst of catastrophe and if climate change enters a vortex of ever-increasing destructiveness, then democratic political institutions could crumble along with national infrastructures.

The chapters that follow examine some of the potential and real impacts of climate change, the political impasse that prevents decisive action on the issue and some solutions for our planet. But first, we take a brief look at the ozone hole.

1 José Lutzenberger, 'Gaia's fever', *The Ecologist*, March/April 1999. **2** Dave Mussell, Juleta Severson-Baker and Tracey Diggins, *Climate Change: Awareness and Action* (Pembina Institute for Appropriate Development 1999). **3** Peter Bunyard, 'Eradicating Amazon rainforests will wreak havoc on climate', *The Ecologist*, March/April 1999. **4** UNEP and WMO, 'Are human activities contributing to climate change?' www.gcrio.org/ipcc/qa/0.3.html **5** As reported by the BBC on 14 March 2006. **6** Steve Connor, 'Global warming "30 times quicker than it used to be"', in *The Independent*, 17 February 2006. **7** The Hadley Center for Climate Prediction and Research, *Climate Change and the Greenhouse Effect*, December 2005. **8** Jeremy Leggett, *The Carbon War* (Penguin, 1999). **9** The Hadley Center for Climate Prediction and Research, 'Climate change and its impacts: Stabilization of carbon dioxide in the atmosphere', October 1999. **10** Grover Foley, 'The threat of rising seas', *The Ecologist*, March/April 1999. **11** Fred Pearce, 'Meltdown in the mountains', *The Independent*, 31 March, 2000. **12** International Federation of Red Cross and Red Crescent Societies, *World Disasters Report 1999* (Geneva, 1999).

2 The role of ozone

A contradictory gas... how the ozone hole affects our health... measures to protect the ozone layer, and how global warming comes into the picture.

THE INHABITANTS OF Punta Arenas in Chile became prisoners in their own homes in October 2000. They weren't hemmed in by insurrection or the threat of infectious disease, but by the sun shining above them. A month earlier NASA scientists had announced the largest-ever hole in the ozone layer above Antarctica, over an area more than three times the size of the US. By October it had extended over the tips of Chile and Argentina, opening up above Punta Arenas. As harmful Ultraviolet-B (UV-B) radiation spread over their city, those venturing out during daylight hours risked irreversible damage to their skin and eyes, which could result in cancer and cataracts. In neighboring Argentina, citizens of Ushuaia were warned that unprotected skin could burn after just seven minutes' exposure. The filter of ozone high up in the stratosphere had been whisked off and the ultraviolet radiation was pouring in through the hole.

By 2003 the record for the largest ozone hole over the Southern polar region had once again been broken and the 2005 hole matched this new record. In Punta Arenas the media provide daily reports on UV radiation levels – a yellow alert stands for moderate levels, but orange or red alerts mean danger. Ozone levels drop as much as 70 per cent; however drops of 10 per cent have been linked to serious health consequences. The health effects are not so apparent in Punta Arenas, where, due to the cold, people tend to go out well wrapped up for only short spells and thus are not allowing the radiation to do its damage. In the country's capital Santiago, 1,370 miles (2,200 kilometers) further north, the ozone levels may not be not as dramatically low, but they affect more lives as people go about their business without perceiving the radiation threat.

What's eating the ozone layer?

Concern about ozone has been around since the 1970s when it was revealed that certain chemicals used in manufactured products and agriculture could eventually start eating into stratospheric ozone 15 miles (25 kilometers) above the earth's surface. These included chlorofluorocarbons (CFCs) and other chlorinated substances used in aerosol sprays, refrigeration and air-conditioning units, and types of artificial foam. Bromine atoms, released by halons (used in fire-extinguishing equipment), and methyl bromide (a pesticide) have the same effect. However, the destruction of ozone is not entirely due to manufactured compounds. For example, chloromethane – which also depletes ozone – originates from forest fires and rotting wood. Other compounds such as nitrogen oxides are implicated too, but the chief agents of the damage after a series of complex interactions in the atmosphere are chlorine and bromine. One free atom of chlorine in the stratosphere is capable of destroying as many as 100,000 molecules of ozone.[1] Such destruction is not permanent, as ozone is constantly forming by the action of intense sunlight over the Equator. But extensive damage to the ozone layer – its thinning in many places and yawning 'holes' opening up – is not something that gets repaired in a matter of a year or two. The ozone hole over the South Pole is a seasonal phenomenon; it closes over in the winter months. But as long as ozone-destroying substances linger in the atmosphere it will reappear each year.

Contradictions

Ozone behaves in quite contradictory ways and, although widely studied, its contribution to climate change is an area that still has many uncertainties. At the lower atmospheric level – the troposphere – where ozone is a byproduct of traffic pollution, it acts like a greenhouse gas, trapping heat in. But in the stratosphere it actually has a cooling effect, which is lessened when the column of ozone thins or when holes appear. Even though estimates are difficult and can change because of seasonal variations in cover, the UN Environment Program (UNEP) reckons that the greenhouse effect from tropospheric ozone is more than twice that of the cooling effect of stratospheric ozone[2] though this range could vary widely.

According to Drew Shindell, a research scientist at NASA's Goddard Institute for Space Studies, this heat-trapping tropospheric ozone may be responsible to a much larger extent for the meltdown in the Arctic than previously thought. As many of the world's highly industrialized (and thus highly polluting) nations are at relatively high altitudes in the Northern Hemisphere, the ozone produced by them is blown by prevailing winds towards the Arctic Circle. 'Instead of being this tiny player, (ozone) can be more like 30 or 40 or even 50 per cent of the cause of warming that we're seeing in the Arctic now,' said Shindell.[3]

Ozone is contradictory in other ways as well, because whereas up in the stratosphere it actually safeguards the health of living creatures by screening out UV-B radiation, at the tropospheric level it contributes to respiratory disorders, more specifically reducing lung capacity, causing chest pain, throat irritation and coughing. Excessive exposure to tropospheric ozone reduces crop yields and forest growth.

When action was first mooted to limit the production of the manufactured chemicals that were harming the ozone layer, manufacturers of products like aerosol sprays were incredulous that chlorine could eventually float up as high as 15 miles (25 kilometers). The evidence, decades later, has left little of that incredulity around. As CFCs were considered quite innocuous before their effects became known, their production rocketed – from

75,000 tonnes in 1954 to up to 800,000 tonnes in 1974. Concentrations are five times as great since then and the bad news is that CFCs have long lifetimes (CFC-115 hangs around for an estimated 1,700 years). Chlorine in the stratosphere should not exceed two parts per billion

The ozone hole's effect on health

One of the impacts of increased Ultraviolet-B (UV-B) radiation is an increased susceptibility of the immune system, reducing natural defenses against infectious and fungal diseases. In areas where malnutrition and infectious disease is already widespread, the effectiveness of vaccines could be reduced. Cataracts are more prevalent among old and malnourished people in poor countries and this prevalence would rise with increased exposure to UV-B radiation.

When UV-B radiation is absorbed by living cells, essential molecules get damaged including DNA which keeps the cell functioning properly. In humans, apart from impairment of the immune response and cataracts, exposure to UV-B radiation brings with it sunburn, aging skin and skin cancer.

In the Netherlands skin cancers are amongst the most common forms of cancer. Despite public education campaigns and being in a temperate zone, the cultural approval for tanning is very widespread. In Australia the incidence of skin cancer is the highest in the world, despite a greater public awareness of the dangers of the sun's rays. But Australia suffers by its proximity to the Antarctic ozone hole and its sunny climate. For its fairer-skinned inhabitants exposure can occur not necessarily due to choice but during relatively short unguarded spells. The outlook for both countries is not favorable – skin cancers manifest themselves after cumulative exposure, that's why they are commonest amongst the elderly. It is predicted that with the damage to the ozone layer peaking right now, the consequent peak in skin cancers can be expected by 2050. Of course future rises in skin cancer rates will depend on factors that are complex to model such as the extent of ozone depletion, changes in behavior regarding exposure and the age of the population, but one study says non-melanoma skin cancers could rise by up to 50 per cent in the Netherlands and 140 per cent in Australia by 2050.

The World Health Organization gives a worldwide estimate. It claims a 10 per cent decrease in stratospheric ozone would cause an additional 300,000 non-melanoma and 4,500 melanoma skin cancers and between 1.6 and 1.75 million more cases of cataract each year. ■

Sources: Pim Martens, *Health and Climate Change*, (Earthscan, 1998); WHO, *Global Solar UV Index*, Geneva, 2002.

in order to prevent further damage to the ozone layer. It is at least three times over that limit.

Besides cancers and cataracts in humans and other animals, UV-B radiation can also lead to plankton losses in clear seas, reducing their ability to remove carbon dioxide from the atmosphere (so spurring on global warming) and adversely affecting the marine food chain. Overexposure to UV-B radiation can lead to plant stunting and genetic mutations in some crops like corn. UNEP has found that damage to DNA in plants due to each consequent drop in stratospheric ozone levels can be disproportionately large.[4]

The Polar regions, where ozone depletion is at its strongest in springtime, thankfully have little human habitation. But thinning ozone beyond the poles increases the risk of skin cancer especially when coupled with fair skins that offer little natural protection against harmful UV-B. According to UNEP a sustained one-per-cent decrease in stratospheric ozone would result in a two-per-cent increase in skin cancers. In early 2005 severe ozone loss was reported over the Arctic – 50 per cent in the layer where the gas normally reaches its maximum concentration, with an overall loss of 30 per cent. The affected region spread over swathes of Europe reaching as far south as northern Italy.[5]

Negotiating a solution

Talks for controlling ozone-depleting chemicals were started in mid-1980 and by March 1985, governments committed themselves to the Vienna Convention for the Protection of the Ozone Layer. The latter year was when the British Antarctic Survey first reported the ozone hole, which sped up negotiations considerably. A Protocol for reductions of ozone-depleting substances was agreed in Montreal in 1987, subsequently revised in the London and Copenhagen amendments with phase-out dates brought forward and more chemicals added to the list of controlled substances. But soon after the Protocol had been agreed upon, scientific evidence showed its provisions to be inadequate and provided the prompt to further action. A schedule for the complete phasing out of ozone-depleting substances was agreed in 1997 – record time for global diplomacy. Even though some sources claim the consumption of the major ozone-depletors has fallen by 80 per cent, it is by no means a figure that is uncontested.[6]

Successes in the fight to make good the damage to the ozone layer must be viewed with some caution, as the damage is peaking at the present time and scientists predict that it will be decades before the seasonal ozone hole will cease from appearing.

In December 2005 estimates of when the ozone hole over the South Pole would heal were extended from 2040-2050 to 2065.[7] The political fallout from the fight also requires inspection. As Kennedy Graham, director of the Project for the Planetary Interest, writes: 'There is a degree of skepticism over the extent to which enlightened governmental leadership forged the way to success of the ozone regime, as compared with a dawning recognition of commercial advantage on the part of the manufacturers during the critical period of the late 1980s. Once the technical and financial viability of alternative technologies was discerned, opposition from the private sector diminished and governments found new space in which to negotiate. Thus, it is sometimes contended, it has yet to be demonstrated that humanity is capable of turning away from future global disaster when compelling corporate interests are in favor of 'business-as-usual'.[6]

An example of this was when the US delegation to a routine Montreal Protocol meeting in November 2003 astonished the gathering by effectively wishing to undo years of steady progress by their demand for the US to be allowed to increase use of methyl bromide instead of the planned complete phase-out by 2005. The chemical is the fumigant most favored by the agricultural lobby in the US and both methyl bromide manufacturers and agribusiness had donated generously to George W Bush's re-election campaign.[8]

Apart from the destruction of the ozone layer, another compelling reason for phasing out ozone-depleting substances is that they are potent greenhouse gases, some with a global warming potential over ten thousand times that of the equivalent quantities of carbon dioxide. (In fact even their substitutes will need to be curbed as they too result in greenhouse emissions.) The Intergovernmental Panel on Climate Change (IPCC) estimates that with the efforts to phase out ozone-depleting substances, their contribution to global warming has dropped by two-thirds.[9]

Some question whether the scale of the effort to provide the necessary transfer of ozone-friendlier technologies to the countries of the Global South is adequate. The transfer of technology is being funded by a Multilateral Fund, into which Northern countries pay their contributions. To date the Fund has over $2 billion, of which it has disbursed $1.59 billion. However the wealthy nations that were responsible for 85 per cent of the consumption of ozone-depleting substances sold nearly $30 billions' worth of them alone in the decade spanning the mid-1980s to the mid-1990s.[10] Technology transfer is also not the clean, efficient process the term evokes; in reality it is muddied by politics, often paving the way for multinational corporations to corner markets in Majority World countries.

With the ozone hole over Antarctica and severe depletion over the North Pole showing little immediate sign of recovery, an alarming feedback is coming into play. Taking a very visible role are some extraordinary clouds that are appearing at stratospheric levels at the Poles. Tear-shaped, iridescent blue and green, and rimmed with bright pink, these polar clouds owe their origin to the cooling of the stratosphere and have come about, curiously enough, as a result of global warming. As the lower atmosphere warms, the stratosphere registers a corresponding cooling. As temperatures drop to their lowest during the sunless Polar winters, tiny traces of moisture in the dry stratosphere get shaped into clouds. Congregating within the clouds are chlorine compounds derived from CFCs. With the arrival of the Spring sun these compounds get the solar trigger, unleashing the unstable form of free chlorine, which then proceeds to eat its way through the mantle of ozone during the polar summer.

It is just such previously unknown interactions, marvelous in their synergy and frightening in their consequences, which are the stuff of global warming.

1 All ozone statistics, unless otherwise indicated, are from Peter Bunyard, *The Breakdown of Climate* (Floris Books, 1999) and Peter Bunyard, 'How Ozone-Depletion Increases Global Warming', *The Ecologist*, March/April 1999. **2** See www.gcrio.org/ipcc/qa/0.4.html **3** Deborah Zabarenko, 'Northern Ozone Pollution Spurs Arctic Warming – NASA', Planet Ark (www.planetark.org), 15 March 2006. **4** UNEP, 'Environmental effects of ozone depletion: Interim summary', September 1999, www.gcrio.org/ozone/unep1999summary.html **5** WMO Press Release 25 April 2005, www.wmo.ch/web/arep/ozone.html **6** Kennedy Graham, 'Ozone protection: Introduction', *The Planetary Interest* ed. by Kennedy Graham, (UCL Press, 1999). **7** Alicia Chang, 'Ozone Hole Recovery May Take Longer', Associated Press, 6 December 2006. **8** Kathryn Schulz, 'The Loophole in the Ozone Layer', *Grist*, 20 November 2003, www.grist.org/comments/gist/2003/11/20/ozone/ **9** IPCC, WMO & UNEP, 'Safeguarding the Ozone Layer and the Global Climate System – Summary for Policymakers and Technical Summary', 2005; available at www.ipcc.ch **10** www.multilateralfund.org (as of 30 January 2006) and Dante Caputo, 'Ozone protection: Argentina', *The Planetary Interest*, ed. Kennedy Graham, (UCL Press, 1999).

3 Impacts: human health

Changes in the spread of disease... a field day for vectors... the unequal battle against ill health... and how epidemics are fueled by weather extremes.

IN CENTRAL INDIA where I grew up, the temperature has hovered for days between 113° and 120° F (45-49° C) in recent summers. While everyone waits for relief from the monsoon rains, people suffer and even die from the heat. This is at its fiercest during the afternoon, with the deadly dusty *loo* wind blowing down the streets. My memory of this seasonal wind is that it would suck moisture right out of the body, drying sweat in an instant, leaving a film of salt on the skin.

But the current heat waves are in a league of their own and they affect the poorest people, those without proper shelter or recourse to fans and adequate water, the worst. In parts of India, unheard of temperatures above 122° F (50° C) have been recorded and the number of deaths from heat that were previously recorded over an entire summer season now occur in just one week. In the 2003 heat wave, over 1,600 people succumbed. In 2005, four men collapsed and died on a single day on a blazing railway platform in the town of Kanpur.

Heatwaves are becoming more common in other parts of the world as well. The one that gripped Europe during the summer of 2003 broke all records for heat-induced fatalities, claiming an estimated 35,000 lives.

The human body is conditioned to patterns of work and rest – overwork or uninterrupted rest are both likely to make us sick. But when just coping with heat becomes a major task for the body, respite becomes increasingly essential to recharge our batteries. As the atmosphere heats up bringing with it stronger heat waves, night-time cooling becomes essential.

Unfortunately atmospheric heating is not uniform and the greatest rises in temperatures are being measured at night (as also in latitudes higher than 50 degrees). The World Meteorological Organization predicted in 2000 a doubling of heat-related deaths in the world's cities by 2020. In urban areas the prolonged heat can be expected to bring clouds of smog and allergens, resulting in respiratory problems.

But heat waves are only a small part of the picture in a warming world. Instability in weather patterns is a marked indicator of warming. In some cases this instability manifests itself as extreme changeability, and in others in the form of more prolonged and intense droughts and downpours. Such disasters not only rack up the death toll by themselves as people get swept away or starve, but they also provide springboards for waves of infectious diseases which, once entrenched, defy eradication. In many cases the aftermath of environmental catastrophe in the developing world is a concentration of displaced people living in camps with little access to safe water and sanitation.

Diseases conquer new ground

However, such threats to human life get little coverage in the Western media, which usually contents itself with occasional rumbles of concern over whether malaria will invade Europe. In the middle of what should be autumn in Rotterdam (where I now live), my sleep is still plagued at night by mosquitoes that hang around longer than usual as temperatures refuse to dip low enough to banish them.

Certainly the risk of malaria transmission is doubled over the coming 20 years for most of Southern and Central Europe if one only goes by predicted temperature changes. Researchers at Durham University have warned of malaria in Britain within the next few decades if current temperature trends persist. However, that is to overlook the sophisticated monitoring and eradication capacities of the wealthy nations in this region. Another story of how mosquito-borne disease has already emerged at higher altitudes in the Majority World due to changing temperatures doesn't grab the headlines in quite the same way.

In September 1999, front pages showed helicopters spraying New York with the pesticide malathion as a response to an outbreak of what was first thought to be St Louis Encephalitis, but turned out to be West Nile Virus. Here was a story that seemed to fit the bill: an 'alien' disease striking at the heart of a grand metropolis and the heroic response to it by the world's wealthiest nation. The link between this mosquito-borne disease and that July's heat wave was easy for most people to make. The sequence of events leading up to the outbreak had started much earlier though. A mild winter had allowed more mosquitoes to survive than was usual and a dry springtime killed predatory insects and concentrated water pools with organic matter, making for rich breeding grounds. The July heatwave provided the springboard. Even though just how the virus arrived in New York remains a mystery, the role the swollen mosquito population played in its spread is crystal clear. In the summer of 2000 the virus was back, identified in mosquitoes in New York's Central Park, and along with it came the helicopters spraying the city. By 2006 mosquitoes had spread the disease to 39 US states and to more than 230 species – not just to humans.[1]

Climate-related disease outbreaks usually occur when a whole string of conditions is right (or wrong, as the case may be) and the weather pulls the trigger. In 1993 a more baffling disease had hit the headlines. People in the US Southwest were coming down with a bug that advanced quickly from a flu-like fever to headaches, nausea and joint pains. But its primary symptom was difficulty in breathing, which progressed to a lethal inability to do so as fluid built up in the lungs. This was *hantavirus* pulmonary syndrome and its carrier was the humble deer mouse, a small rodent.

The virus exists either inactive or isolated in small rodent populations without much effect on humans, but when the conditions are right for a population explosion the story is quite different. In 1993 a prolonged spell of disturbed weather brought with it just those conditions. First a drought cut down the populations of the deer mice's enemies – birds of prey, coyotes and snakes. Then heavy rains early in the year brought plentiful grasshoppers and nuts for the mice to feed on. The mice wasted no time in breeding, setting off a sudden increase in the carriers of the virus. By summer drought conditions were back and the swollen ranks of deer mice left the open spaces to forage for food where humans lived. This close proximity brought what was then a mystery disease onto people's television screens.

In the same way that tiny particles of shed human skin can float around as motes of dust in our homes, particles of infected rodent urine, saliva and droppings got into the air in the form of a fine mist that was then inhaled. This was all it took to transmit this life-threatening disease. By autumn the deer mouse population had subsided and the outbreak came to an end. Later researchers would have to don head-to-foot protective clothing, looking a bit like astronauts, in order to study the disease in sealed deer mice pens. Today the US has early-warning systems in place and rodent populations are closely monitored, but the disease has emerged in Latin America, where such sophisticated methods of control may not be financially viable.

Borne on a buzz

Mosquitoes play the most significant role in projections of the climate-change induced spread of disease. They can relay malaria, dengue fever, yellow fever and many kinds of encephalitis. With growing drug-resistance and the decline in public health efforts in many developing countries, malaria today claims the lives of at least one million people and causes over 400 million episodes of serious illness each year. A disease that causes debilitating fever cycles that leave the sufferer feeling weak and sapped of strength, it has quadrupled between 1995 and 2000.[2] Its re-emergence in parts of Southern Europe and Russia, the Korean peninsula and the Indian Ocean coast of South Africa may be due to a range of environmental and human factors. But its emergence in highland areas can be more directly attributable to global warming. Some climate models foresee the zone of the disease encompassing 60 per cent of the world's people by the end of this century, adding an extra 50 to 80 million cases of malaria per year.[3]

The changing climate can influence the transmission of malaria in different ways. Prolonged warm periods

Increased range

Many highland regions in the tropic and temperate zones, previously too cold for mosquitoes, now provide hospitable temperatures for these disease-bearers. In the tropics, elevations at which temperatures are constantly freezing have climbed a full 500 feet in the past 30 years. Here's where the *Aedes aegypti* mosquitoes and mosquito-borne disease have emerged recently:

Malaria (spread by *Anopheles* mosquito)
Highlands of Ethiopia, Rwanda, Kenya, Uganda and Zimbabwe,
Usambara Mountains, Tanzania
Highlands of Papua New Guinea and West Papua

Dengue fever
San Jose, Costa Rica
Taxco, Mexico

***Aedes aegypti* mosquitoes**
(which spread both dengue and yellow fever)
Eastern Andes Mountains, Colombia, Northern highlands of India, Highlands of Uganda, Ethiopia, Kenya and Rwanda.

Scientific American, August 2000; *The Heat is On*, Ross Gelbspan, Perseus Books, 1998.

above 60° F (16° C) are a precondition for *Anopheles* mosquitoes to transmit the most virulent malaria parasite *Plasmodium falciparum*. As it gets warmer, mosquitoes breed rapidly and bite more often. The parasite itself halves the time in which it develops fully in the mosquito's body during hot spells, doubling the chance that it will be transmitted before the mosquito dies. But other intense weather phenomena such as flooding and droughts can play a role too – with pools being left behind as floods recede and streams becoming stagnant during droughts.

Also broadening its range, piggybacking on *Aedes aegypti* mosquitoes, is dengue fever which prefers temperatures that do not often fall below 50° F (10° C). No accurate treatment for it exists and its nickname – breakbone fever – gives some indication of the severe pain associated with it. People who have contracted dengue liken it to arthritic pain, but far worse – as if their bones would snap. In its hemorrhagic form dengue can be fatal. Up to 100 million people contract dengue fever each year and in the 1990s its range had widened in the Americas and crept up to northern Australia.[4] Previously *Aedes aegypti* were confined to altitudes below 3,000 feet (1,000 meters), but now they have been found as high up as 6,600 feet (2,200 meters) in Colombia.[5]

Further increases in the latitudinal and altitudinal range of dengue and a longer transmission season in temperate regions are on the cards.[6]

Disease in a world divided

The impact of climate change on human health is difficult to quantify. This is because there is no easy correlation between, say, rising temperatures, increase in vectors and spread of disease. There is a host of social and economic factors which might make all the difference when it comes to whether a certain combination of conditions will result in an outbreak or not. In 2003, the World Health Organization estimated that 150,000 deaths were caused by climate change in 2000 with a further 5.5 million years of healthy life lost due to related disease. They predict a doubling of this in the next 30 years if current trends are allowed to follow their course.[7] The IPCC puts

A plague on all your houses

Historian ***David Keys,*** *author of* Catastrophe: an Investigation into the Origins of the Modern World, *gives his view of the plague epidemic that swept the 6th century – and his belief that its origins lay in climate chaos.*

'With some people it began in the head, made the eyes bloody and the face swollen, descended to the throat and then removed them from Mankind. With others, there was a flowing of the bowels. Some came out in buboes [pus-filled swellings] which gave rise to great fevers, and they would die two or three days later with their minds in the same state as those who had suffered nothing and with their bodies still robust. Others lost their senses before dying. Malignant pustules erupted and did away with them. Sometimes people were afflicted once or twice and then recovered, only to fall victim a third time and then succumb.'

Thus wrote the 6th-century church historian Evagrius, describing the gruesome symptoms of the bubonic plague that devastated the Roman Empire and much of the wider world in the 6th and 7th centuries, the so-called Dark Ages.

By decimating populations and wrecking economies, the plague transformed the history of the eastern Mediterranean, Western Europe, the Middle East and Africa. In a sense, it was the final nail in the coffin of the Classical World and helped give birth to a string of modern European nations – England, Ireland, France and Spain. But the plague was not simply a disaster that erupted out of nowhere. My research shows that it was almost certainly triggered by the climatic chaos of the mid-sixth century AD. It seems that the disease had long been endemic among wild rodent populations in East Africa – but it was the climatic disaster following 535 AD that enabled the disease to spread outside its normal territory.

The climatic situation (probably cooler, drier weather followed by floods) helped the plague bacillus in three key ways. Cooler weather increased the population of the fleas which carried the bacillus in their gut. It also forced the fleas to bite more rodents and other mammals because cooler temperatures prevented the bacillus releasing a natural anticoagulant in the flea's gut, a failure that resulted in the flea becoming ravenously hungry as its gut was blocked by blood clots. Third, the climatic chaos destabilized the relationship between predators and their rodent prey – a destabilization which led to a breeding explosion by the flea's rodent hosts. Once the disease had broken out of its normal territory, it appears to have spread throughout much of it characteristically mildly: 'Overall, negative health impacts are anticipated to outweigh positive health impacts.'[6]

One of the ways of combating the spread of disease in a warming world would be to take intensified adaptive steps. This would mean improving surveillance systems

Africa – and northwards into the Mediterranean world. Its journey to Europe and the Middle East was by way of the Red Sea and the Roman equivalent of the Suez Canal – and was almost certainly transmitted courtesy of Roman greed for African elephant ivory. It was probably the ivory-trade vessels that introduced bubonic plague into the Mediterranean world.

I believe that in Africa it killed off the continent's major ancient ports. Indeed to this day their precise locations are a complete mystery to archaeologists. The plague also fundamentally changed the nature of the culture and the economy across vast swathes of eastern and southern Africa. Agriculture declined in many areas and pastoralism took over – almost certainly because plague-carrying rodents were more attracted to agricultural settlements and their stores of grain than they were to milk and meat on the hoof.

...and one recent instance of re-emergence

In David Keys' account above, plague spread unchecked across the continents. It is the knowledge of the devastation wreaked by plague epidemics in the past that spread fear in Western India during the rainy season of 1994.

The outbreak began in the merchant city of Surat in Gujarat state. That summer temperatures had been sweltering and when the rains came they stayed longer than usual. The raised levels of humidity proved the trigger for a flea explosion in the city's grain-storage depots. As the rains continued, the drainage system became overwhelmed. As flood-waters rolled over the city streets, garbage bobbed up on them and spread out. Rats, surprised by the feast, lost no time in multiplying. After this it was the classic recipe of rats, fleas and eventual human infection, which then spread to hundreds of people. The news sent shockwaves through the country and beyond, making headlines in international news bulletins. Fears that people who had been infected might have traveled to the densely populated metropolis Mumbai (Bombay) sent the city's rat catchers onto red alert. There was talk of people not being allowed onto international flights from the area. Airline and hotel businesses lost over $3 billion as flights and hotel bookings got canceled. Meanwhile in Surat the municipal authorities swung into gear and had the streets lined with rat poison. Eventually the outbreak subsided, but not before it had shocked the world into realizing that plague had not gone away and that climatic extremes could wake it up. ■

and, once potential threats were spotted, taking measures to limit the populations of carriers of the disease (such as mosquitoes), advising people on protective measures they could take and providing preventive medicine where possible. Many industrialized nations already do this, while many poorer countries, which find even basic health care provision a struggle, fail. If climate change increases the potential for outbreaks of disease, the fallout would once again be greater in the Majority World.

As it is, recent disasters have taken a heavy human toll – the WHO estimates that 600,000 people lost their lives due to weather-related natural disasters in the 1990s; 95 per cent of them lived in poorer countries.[8] But the number of deaths resulting from diseases following such disasters can often rise much higher. In February 2000 freak cyclones and heavy rains flooded large parts of southern Africa. The TV cameras captured dramatic footage of bodies swept away, but did not stay to witness the aftermath – the spread of malaria and cholera in Mozambique and Madagascar.

In Kenya, unseasonably heavy rains in early 1998 caused a cattle disease called Rift Valley Fever to jump the species barrier and kill more than a thousand people within a few weeks. Such epidemics following natural disasters can stop development in its tracks for years.

As climate change throws up longer-lasting extreme weather and sudden dramatic changes, a rise in waterborne diseases can be expected, especially in regions with a weakened infrastructure. Dysentery and cholera often emerge in refugee camps following relatively short spells of rainfall because people are at their most vulnerable, living close to the elements with no access to a safe water supply or adequate sanitation. But severe weather can also threaten those who are not quite as vulnerable as refugees in camps. Droughts can drive people to unsafe water sources and increase the number of contaminants in the water available. Water scarcity also affects basic hygiene. In the case of floods, sewage and fertilizer can get swept into the drinking water supply, especially when livestock farms are situated close to water channels. These contaminants can cause algae blooms in the water making fish unsafe to eat and providing breeding grounds for cholera.

All of this is nothing new. It is just that a changing climate will tip the ecological balance that keeps many of these diseases at bay more often. The massive cholera epidemic of 1991 that spread from a port town in Peru to Ecuador, Colombia, Chile, Guatemala, Mexico, Panama and Brazil and which infected more than half a million and killed 5,000 is one such fearful example. Where it all began in Chimbote, Peru, unusual weather provided the ideal conditions. Along the coast there were widespread blooms of algae due to warm surface waters, making them a nutrient-rich soup. This was coupled with severe, unseasonal floods which tipped sewage into the water supply and started off the chain of infections.[5] The epidemic cost Peru $1 billion in lost tourist revenue and seafood exports.[9] Each El Niño episode leads to well-recorded disease surges.

With this Pandora's box of sickness in mind, the health benefits of a warming world seem slight. Superhot temperatures in the tropics could reduce the *schistosomiasis*-bearing snail population. Milder winters in colder regions could reduce cold-related heart attacks and breathing problems, but the number of lives thus saved would not be much of an improvement on the numbers lost due to increased heat stress.[4]

Fumes and food

In this assessment of future ills that could result from climate change, the focus has been mainly on the spread of infectious disease. But there is another more direct link with health and human meddling with the climate – one that is evident before climatic changes make their presence felt. The pollution that causes a build-up of greenhouse gases in the atmosphere affects human health too. Cities where skies are obscured by the emissions of industry and exhausts from thousands of wheezing cars – often stuck in traffic jams – have higher rates of asthma and cardio-respiratory disorders. Many climate experts only concern themselves with greenhouse gases' effects on the global climate, ignoring the issue of local pollution, whether it be from toxic pollutants from vehicle exhausts, fine particulates, heavy metals, or others from an ever-lengthening list. Cleaner environments are in everyone's best interests.

The changing climate will also have implications for food production (see chapter 4). The Hadley Center of the British Meteorological Office predicts that, despite regional variations, the negative changes in food production will be most marked in the tropics, with Africa being the worst affected. In Africa a predicted 18 per cent more people will be at risk of hunger due to climate change alone by 2050 and the forecast for tropical South America is also bleak. Needless to add these are regions which can ill afford to sustain such losses in food production, as already shortages are by no means remarkable events.

For subsistence farmers on marginal and vulnerable lands, relatively slight changes in rainfall could have intensely magnified effects, as growing food is a challenge on such lands at the best of times. Cynics would argue that the real reason is that overpopulation drives people to farm on marginal land. However, population is only a part of the picture with dispossession due to poverty and resource-grabbing by the rich playing a more significant role. The Hadley Center calculates that 'the number of people at risk of hunger is projected to increase due to climate change by 30 million by the 2050s.'[10] This estimate does not account for large-scale catastrophic events or a spiraling cycle of positive feedbacks. The threat of malnutrition and the increasing susceptibility to disease it brings is about as basic as you can get. Any food crisis is also a health crisis, as television images of famine remind us.

The health impacts of the deterioration of living conditions caused by rising sea levels, droughts and floods and other freak weather are determined not just by the phenomena themselves and the displacement they cause, but by a complex web of social and economic factors that sing along as a ghastly chorus to the health challenges. When the weather plays havoc with people's ability to work for their keep and harvest local resources, it compounds weakness and disease.

The ultimate health risk could be posed by a sudden threshold in the global climate system caused by runaway feedbacks – making parts of the inhabited world much hotter or even much colder. The likelihood of this happening is fortunately still estimated as being relatively low, but there is no denying that it is probable. In such a scenario the health implications would not so much be about combating disease and hunger, though. Rather it would be a question of coping with the weather itself.

1 http://westnilemaps.usgs.gov/us_human.html **2** Ross Gelbspan, *Boiling Point* (Basic Books, 2005). **3** *World Disasters Report 1999* (International Federation of Red Cross and Red Crescent Societies, 1999). **4** Paul R Epstein, 'Is Global Warming Harmful to Health', *Scientific American*, August 2000. **5** Ross Gelbspan, *The Heat is On,* (Perseus Books, 1998). **6** *Climate Change 2001: Impacts, Adaptation and Vulnerability* (IPCC, 2001) **7** 'WHO Says Climate Change Killing 150,000 a Year', *Reuters*, 11 December 2003. **8** 'Climate and heath fact sheet, July 2005', WHO. **9** 'To save lives, give global warming the same priority as biological weapons, says WWF', World Wildlife Fund/World Wide Fund for Nature press release, 5 November, 1998. **10** 'Climate change and its impacts', The Meteorological Office, London, November 1998.

4 Impacts: farming and food production

Farmers facing adversity... models for a warmer future and their limitations... the threat of rising seas... the problems besetting modern agriculture and how they would be amplified by climate change.

TARA BEGUM WAS used to floods. As a small farmer in the mighty Brahmaputra River's delta in Bangladesh she had to be. The river is renowned for its summer ferocity when the Himalayan glaciers from which it springs release more meltwater. This is also the season when the monsoons bring down water from the heavens and further swell the country's myriad rivers and creeks. Storm surges whipped up by the monsoon winds can also inundate lands. But Tara Begum was used to that – just like other Bangladeshi farmers. They know how to make the best use of floodwaters and to plant their crops at the most appropriate times.

But then the Brahmaputra got fiercer. Tara Begum and her children had to shift the shack in which they lived five times. Eventually her land became permanently flooded. 'The river gobbled up all we had,' she explained. 'We were at its mercy.' This was the result of the accelerated melting of Himalayan glaciers due to global warming.

Tara Begum picked herself up; it was either that or starve. She now depends on a weed that thrives in the waters. Tying up large quantities of water hyacinth into a floating island, she then sprinkles it with a layer of soil and plants her vegetables in it. This is some sort of success story. But, one wonders, for how long?[1]

Farmers on the whole are a canny lot, especially in regions where a single climatic phenomenon like seasonal rains determines their harvest. They are used to adapting what crops they grow judging by the kind of growing season expected. In India if the monsoon is late most farmers have to forego planting their crops of choice, relying on standbys like coarse millet. But if the monsoon is severely delayed, then the choice is whittled down further to much lower yielding options.

The monsoon is behaving erratically over this vast subcontinent and millions of farmers are paying the price.

Take Chimanbhai Parmar from India's Gujarat state. The farmers in Parmar's village have been trying to get used to the fact that the monsoon, once a regular visitor in June, has in the last decade been leading them quite a dance. One year with the arrival of the first showers more or less on time, Parmar and the farmers of his village praised the rain gods and set about planting peanuts, their only cash crop. But the wait for the next set of showers, which usually follow in regular pulses, seemed to stretch out endlessly. Within a month of sowing, the plants had paled and withered in the scorching sun. Two months after sowing, his fields lay parched and barren. For miles around the same devastation was in evidence. Parmar, with three children to support, joined his fellow farmers in looking for non-agricultural work. Used as they were to hardship, their options were still unattractive to say the least – piecemeal work in local shops and factories or daily-wage labor for the Public Works Department, at less than a dollar a day.[2]

In my home state of Madhya Pradesh many recent monsoon seasons have seen rainfall far below the norm. The mercury shoots up, saplings lose the fight, and food becomes scarcer and more expensive. Temperatures above 104° F (40° C) can lead to wilting because they damage essential proteins in plants. With the heat-compacted soil turning warm, roots get affected, becoming less able to take up nutrients. Recent summers in Madhya Pradesh have seen the temperatures even go over 122° F (50° C). This has left even people who are used to the baking summer in this region bewildered.

Meanwhile Bangladesh's floods become more and more violent. In 2000 torrential downpours led to the loss of over 1,200 lives and rendered three million people homeless. When humans found it hard to survive the weather's onslaught, crops had little chance.

Many African countries seem to go from drought to drought. In Africa the failure of the rains is less unusual than in Asia, but for the rains to fail year on year signifies something is seriously amiss.

The outlook from climate models

Such tragedies put into sharper focus the shifts and changes in world agriculture predicted by the climate-modelers of the IPCC. While the scientists talk in terms of general trends and percentages, the situation for the farmer on the ground can sometimes vary from one field to the next.

On the subject of water scarcity their conclusions are quite frankly frightening. The IPCC concluded in its 2001 report that the number of people experiencing water stress (currently estimated at 1.7 billion) would increase to 5 billion as soon as 2025, depending on the rate of population growth.[3] This must be set against the shortages that are already in evidence in many parts of the world today and have become highly politicized issues. So far, so bad.

When making predictions about food supply the unifying thread is that the lower latitudes, in particular the arid and semi-arid regions of the tropics (where agriculture already poses great challenges), are likely to witness drops in yields and increased risks of hunger. Increased carbon dioxide in the atmosphere could boost plant growth in some temperate regions, especially for cereals such as wheat, by speeding up the photosynthesis process by which plants make food. Such predictions have been eagerly seized upon by the fossil fuels lobby, which has promoted the idea of a future agricultural cornucopia with vigor.

But don't hold your breath. The scientists are not quite so enthusiastic: 'Many of these potential benefits are very small and are not significantly different from those arising due to natural decade-to-decade variability.'[4] And there is the additional fact that not all food crops benefit from higher levels of carbon dioxide – corn/maize, sorghum, millet and sugarcane are some that lose out – whilst many aggressive weeds do. An increase in carbon dioxide would force plants to respire more rapidly. In hotter areas, such as Southeast Asia and India, a mere 0.9° F (0.5° C) of further warming would lead to steep falls in rice and wheat production.[5] To say nothing of the explosion of crop pests that could occur (see box: Very, very hungry). Among the major losers in all scenarios are Africa, a continent where food production is already struggling to keep pace with population, and India, the world's second most populous

Very, very hungry

Despite the use annually of millions of tons of pesticides, more than 40 per cent of the world's food crop is lost to pests, plant diseases and the effects of weeds – a yearly loss estimated at $500 billion.

Warmer temperatures and milder winters will offer improved opportunities to pests. They will be able to spread to higher latitudes and altitudes and benefit from a longer active period each year. Larvae may survive over winter in areas where they are killed off by the cold and cause major infestations when the next cropping season arrives. Insect pests are capable of producing numerous generations in a year with the females of some species producing hundreds of offspring. The kind of critical mass that locusts can generate – capable of stripping every field in their flight path – could become a reality for Southern Europe. A rise in temperature of 1.8° F (1° C) would enable the European corn-borer to chomp its way 300 or so miles (500 kilometers) northwards. A 5.4° F (3° C) rise would see a profusion of insect pests, which would be difficult to control no matter how many chemical poisons one might fling at them. Animal diseases such as African swine fever could migrate to North America. Plant diseases, especially of the fungal and bacterial variety, would luxuriate in warmer, wetter conditions. In Britain the milder winters are already prompting outbreaks of potato blight and mildew in cereals. ■

Sources: Peter Bunyard, 'A Hungrier World', *The Ecologist*, March/April 1999; Cynthia Rosenzweig and Daniel Hillel, 'Potential Impacts of Climate Change on Agriculture and Food Supply', 1995, www.gcrio.org/CONSEQUENCES/summer95/agriculture.htm.

nation. The US, the world's leading grain producer, and Canada could expect net gains under a modest warming scenario, but this soon declines with further warming to a net loss. Recent field experiments conducted by the University of Illinois found the increased yields due to elevated levels of carbon dioxide were only half of those predicted and when the rising levels of tropospheric ozone in the Northern Hemisphere were factored in yields actually decreased.[6]

In 2005, Martin Parry of the UK Meteorological Office's Hadley Center told a British Association science conference: 'We expect climate change to aggravate current problems of the number of millions of people at risk of hunger, probably to the tune of 50 million [by 2050]. The greatest proportion, about three-quarters of that number, will be in Africa.'[6] Previous assessments by the Hadley Center had estimated the number at 30 million additional people. As the models improve, so do predictions of damage to come. Some other predictions have ranged in the hundreds of millions. It becomes difficult to remember that these numbers will translate into individual people, several hundred sports stadiums full, facing chronic hunger and slow starvation.

Such global averages cannot hope to reproduce the necessary level of detail that would be required to give an accurate prediction for each region. One Hadley Center report acknowledges as much: 'Note that this is a global and long-term assessment, focusing on average effects over space and time. At the local level (for example, in especially vulnerable areas) and over short periods (for example, in spells of drought or flooding) many of the effects of climate change on agriculture will be more adverse.'[4] The devil, as they say, is in the detail.

With established patterns of weather being disrupted more frequently, it would matter little if the average annual rainfall or temperature evened out in the end – it's the extremes that lurk within those averages that would damage the production of food. Also averages over relatively large geographical areas would be misleading if conditions in smaller food growing belts had been less than ideal. There is also the important consideration that all farming is dependent not just on rainfall but when that rainfall occurs – for example a prolonged hot and dry spell when the crop is coming into flower and the fruit forming can dramatically reduce yields.

The sea and salinity

Agriculture in proximity to densely populated coastlines would suffer from rising sea levels. Seas will rise by half a meter by 2080 purely on account of the fact that warmer water takes up more volume. But if we add on the probability of significant melting of ice sheets in Greenland and the potential destabilization of the Western Antarctic Ice Sheet – the rise would be considerably higher. A 3-foot (1-meter) rise would wipe-out to a full third of the world's croplands. In low lying coastal areas where drainage is already a problem – among them parts of Egypt, Bangladesh, Indonesia, China, the Netherlands, Florida – it would be an uphill battle to sustain agriculture.[7] The sea's assault would be two-pronged: by creeping increases in salinity in soil and aquifers near the coast; and by coastal flooding and storm surges. The experience of countries like Bangladesh in the recent past has been of surges raising a wall of water up to 18 feet (6 meters) high. One estimate suggests that in the next 50 years, surges in Bangladesh could cover up to 40 per cent of the country's land.[8] The scale of the environmental, refugee and food scarcity problems that would result is difficult to fathom.

Tropical mangroves that protect against the infringement of sea water into croplands have been chopped down in order to raise shrimp destined mainly for consumers in industrialized countries. In Thailand and Bangladesh, shrimp cultivation has led to the salinization of rice fields further inland. With the salt goes the crop and the land becomes barren. Sea water also penetrates where there is a gap created by the over-extraction of water from aquifers. In Egypt salinity has moved 22 miles (35 kilometers) up the Nile Delta[8] and the future for the country's farmers looks daunting with the prospect of coastal flooding and a drying up of the interior as a result of hotter temperatures.

In the constant search for greater productivity, irrigation has both answered prayers and given cause for new ones. Enabling farmers to grow up to three crops a year, irrigation accounts for 40 per cent of the world's food. But in the Majority World it also brings with it problems of salinity and saturation. So with old lands becoming unproductive, new fields must be found, usually at the expense of forests. Irrigation often relies on tapping water from deep underground aquifers, but many countries including the US have not been successful in drawing on this resource at a rate that would allow for replenishment. Grain and cotton

Continental breakdown

The British Meteorological Office predicts that by the end of the century a third of the globe's land will be subject to extreme drought – a sevenfold increase. This despite the general prediction of increased rainfall globally caused by heightened levels of evaporation. Welcome to a world of weather extremes.

Here are the IPCC's main predictions related to agriculture at a glance:

Africa
- Grain yields drop.
- River flows reduce for Mediterranean and southern countries of Africa.
- Increases in droughts, floods and other extreme events.
- Desertification intensifies due to lower rainfall, washing away of topsoil due to storms and drying out in hot spells.

Asia
- Droughts, floods and storm cycles have already increased.
- Excluding northern areas, a general decline in farming productivity due to excessive heat and water stress, sea level rise, and extreme weather.
- Sea level rise and cyclones would lead to tens of millions abandoning coastal areas, threatening the biodiversity of interior parts.

farmers in the southern Great Plains continue to deplete the massive Ogallala aquifer.

It's a similar story in aquifers in California's Central Valley, home of half of the US's fruit and vegetable produce. In the central states of India, the advent of tube wells seemed like the answer to the variability of the monsoon. Whereas traditional wells often ran dry during the wait for the monsoon rains, tube wells guaranteed a steady supply of water. Or they did until recently. Now water tables in many regions have dipped alarmingly and the tube wells are running dry. Efforts to sink them ever deeper bring little reward. Non-governmental organizations (NGOs) have embarked on massive education programs, encouraging farmers to return to traditional water harvesting methods, such as saving water when it falls. Such methods, once widely employed and part of Indian farmers' environmentally-friendly ways of harnessing limited resources, had fallen

Australia and New Zealand/Aotearoa

- An initial benefit to temperate crops turns to losses as climate change progresses.
- Drying trends over the entire region.
- Flooding, storm surges and wind damage due to more intense rains and cyclones.

Europe

- Summer water shortages in Southern Europe; winter rainfall increases for both north and south.
- Agricultural productivity decreases in Southern and Eastern Europe.

South America

- Floods and droughts become more frequent; flooding degrades water quality.
- Crop yields decline; subsistence farming in some regions threatened.
- Biodiversity loss speeds up.

North America

- Droughts in Canada's prairies and the US Great Plains.
- Northern areas become more accessible to farming.
- As warming progresses, benefits to crop yields turn to net losses.

Source: IPCC, *Climate Change 2001: Impacts, Adaptation, and Vulnerability*. www.ipcc.ch

into disuse, ironically enough, with the promise of piped water and tube wells.

Revolution gone sour

The challenges our world – with its increasingly voracious North and a more frugal but increasingly populous South – faces with respect to food production are already quite daunting. Within a time frame of 30 years we are looking at doubling harvests, otherwise there could be shortages resulting in chaos. While innovative schemes of sustainable agriculture are springing up in many of the poorer countries of the world particularly in Asia, as are communal efforts at growing and sharing food, the ravages of free market economics are also biting deep. For many farmers in the Global South, cash cropping is the only alternative to joining the ranks of the dispossessed in urban slums. There, as elsewhere, the Green Revolution – which ultimately mainly benefited richer farmers and transnationals that want a say in everything from what

What farming contributes to the global greenhouse

Farmers release greenhouse gases in many ways. Carbon dioxide is released by clearing forests for crops, burning farm waste, and in the case of industrial agriculture, by use of fossil fuel energy to move heavy machinery and the production processes involved in the making of pesticides and fertilizers. Methane is released from rice paddies and large herds of livestock. Nitrogen oxide comes mainly from nitrogen-based fertilizers. The Intergovernmental Panel on Climate Change estimates that agriculture is responsible for 20 per cent of greenhouse gas emissions – via activities such as soil fertilization, factory farming of animals and rice farming. A further 14 per cent is added by land-use changes, such as the clearing and burning of vegetation, often to make land available for farming. ■

Sources: Cynthia Rosenzweig and Daniel Hillel, 'Potential Impacts of Climate Change on Agriculture and Food Supply,'1995, www.gcrio.org/CONSEQUENCES/summer95/agriculture.html; Nick Sundt, 'Agriculture and Climate Change: a hard row to hoe,' July 2000, www.globalchange.org/featall/2000winter2.htm

is grown to what fertilizers and pesticides are used and what price they are willing to pay for the end-product – has had less than green results. Originally intended as a short-term fix for farmers in countries with agricultural deficits, it birthed a kind of industrial farming that brings soil erosion, increased pesticide and fossil fuel use and water pollution in its wake.

The problem of soil erosion is particularly acute, as one inch (2.5 cm) of topsoil takes around 500 years to form and is now being eroded at a rate that outstrips its replenishment. The loss in terms of nutrients has been estimated at close to the amount of fertilizer being used in the world's agriculture today. With higher temperatures heating up the soil, nutrient loss will speed up and the topsoil will get compacted and more prone to being washed away. We have seen already that the advance of desertification in marginal lands has been abetted by prolonged droughts (see box: A desert is born). When downpours follow on such droughts you can see the topsoil floating away in front of your eyes in the murky rush of floodwater.

Traditional methods of farming in India which emphasized subsistence crops over cash crops and which depended upon human and animal labor were derided in the heady early days of the Green Revolution. Instead a vision of modernity was pushed at an international level and promoted by an Indian government aspiring to the success of the rich West. The cattle-drawn plow was replaced by the carbon dioxide spewing tractor, and dung used as fertilizer gave way to chemical fertilizers that released nitrous oxide into the atmosphere.

Such systems of farming necessarily benefited richer Indian farmers at the expense of the much more numerous poor. Indian environmentalist Vandana Shiva spelled out the implications way back in 1990: 'There is nothing inherently productive about agriculture based on oil and chemicals. In India traditional farmers use about half a calorie of clean renewable energy to produce one calorie of food, while highly mechanized, chemically-based agriculture uses ten calories of polluting, nonrenewable energy to produce one calorie of food.'[9] She had little time for lofty claims that this was 'helping' agriculture in countries like India: 'Aid for fertilizer, tractors, transport and energy mega-projects: all of these have primarily been ways for Western corporations to sell more machinery, equipment and engineering services to the Third World. For every dollar of aid given, three dollars' worth of business is generated in the industrialized countries.'

Shortages and inequity

On top of such ills that beset modern agriculture there is the naked fact of waste and want. In the US a government study revealed that more than a quarter of all food produced in the country doesn't get eaten,[10] while in the UK supermarket food capable of feeding 270,000 people ends up being trashed each year. With money-economies firmly in place, something as basic as food production is completely skewed by the demands of wealth. Industrial countries with about a quarter of the world's population, manage to use up about half of the world's grain, often feeding it to animals that will be eaten as meat.

Perhaps this would not matter so much if the rich countries were self-sufficient themselves – but they are not. They corner resources through the sheer power of money. This is where arguments about 'population control' in the global South as a means of conserving resources and preventing environmental degradation seem to miss the point. If one starts from the premise upon which most declarations of rights are based, that every person is equal, then it should follow that each person has the same right to resources as everyone else. But if we look at the consumption of resources we see that the reality is vastly different. If population were measured in terms of the amount consumed, then the United States would have a population equivalent to twice that of China and nearly six times as large as India – the two most populous countries in the world.[11] Perhaps it would be wiser first to address such overconsumption, before 'overpopulation'.

While many poor people in the Majority World work in conditions akin to slavery, producing crops for the world's food giants for a pittance and while speculators make their fortunes on the futures of these commodities, institutions like the World Bank insist on a 'level playing field' – that is, the removal of all price supports which might bring these workers something more than a starvation wage.

If one then factors in extreme weather events and the possibility of more positive impacts in the richer North while the South suffers most of the fallout of the weather we are left not only with a vision as Argentinean diplomat Raul Estrada Oyuela put it of 'a green North and a brown South', but of a South where all forms of instability would be the order of the day. Climate change optimists take the line that the possible increased food production in the countries of the temperate latitudes could offset losses elsewhere – as if this excess food would just be given away to the disadvantaged. Environmental

Water, wheat and beef

All farming needs water. But the amount of water needed to produce a pound of beef is far greater than that required for a pound of wheat.

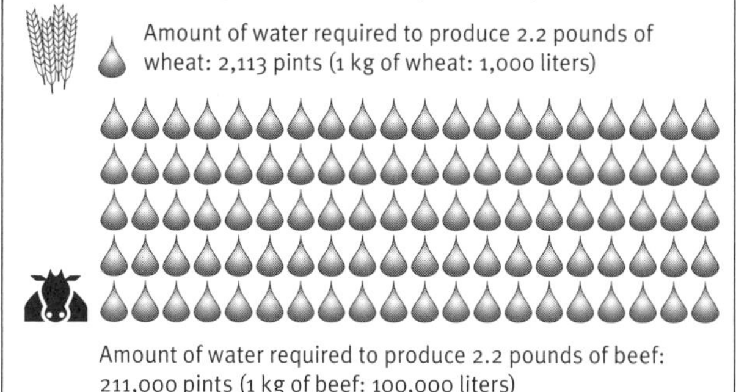

journalist Ross Gelbspan writes about the possible consequences of such a divide: 'The greenery of such a North would be deceptive. It would conceal a political and moral time bomb. It is hard to imagine that a society that fortresses itself against the rest of the world could continue to be an open society, vibrant with freedom, productively democratic, peaceful, and secure.'[5] Many Southern commentators would argue though that the economic stranglehold the North already has on their nations now calls into question any notions of an 'open society'.

Changes in the northern latitudes

But whether the outlook will really be quite so cozy for countries in the temperate latitudes is hotly contested and there are arguments aplenty why it may instead be deleterious. The US is currently responsible for nearly half the world's exports of grain. But since the 1980s droughts have been an increasingly frequent visitor in the US heartlands. The 1999 heat wave that killed over 270 people and thousands of fish also hit farmers hard. A $7.4 billion aid package was required to support them. After a record warm winter in March 2000, the US Agriculture Secretary said: 'We saw last summer just what a drought can do to farmers. Looking to the future, we need to be ahead of the curve, prepared for dry weather when it comes and equipped with the mechanisms that will protect farmers and prevent widespread losses.' Well the future arrived just a few months later, when heat waves and drought conditions returned and forests went up in smoke. Since then not a year has gone by without drought being reported in the US.

The long-term future looks likely to see further reprises. Researchers of the National Oceanic and Atmospheric Administration (NOAA) predict that the central US will see 'substantial percentage reductions' in available moisture during the summer season by 2050 – more frequent drought conditions and reductions in yields of crops.[12] Many of the studies that talk of a more favorable outcome for US agriculture's adaptation to climate change assume the availability of water – with this essential assumption knocked out of the equation things begin to look very different indeed. Irrigation requirements would shoot up especially if warmer temperatures accompany the lack of rain and there are fears that the droughts could result in a return of the disastrous Dust Bowl conditions in the US Midwest.

What are the adaptive prospects available to the farmers of wealthy nations? Better management of resources and technological advances are mooted, as are the adoption of drought and pest resistant varieties of crops. The bottom line remains, however, that all such adaptations will require more capital outlay and crops that can grow in drier conditions will have lower yields. Northern Canada and Russia, similarly assumed to be beneficiaries of climate change, could also experience warmer drier summers. If, as models have predicted, the production of crops will become viable in the higher latitudes, several other issues are raised, such as the fact that conversion of land for farming purposes will release more carbon dioxide into the atmosphere when other vegetation is removed and the organic matter in soil disturbed. Soil fertility here is lower than in the traditional croplands and crops that are adapted to the length of day in lower latitudes could react unpredictably to longer daylight hours in the north, possibly with early maturation that would reduce yields.

The outlook for Europe looks similarly dismal. Areas *south* of 48 degrees north will get drier. In June 1999 Spain had the worst drought in 50 years wrecking agricultural production. In 2005 it was Portugal's turn, although Spain and France were affected too. The rest is in for soggier conditions. A study by researchers Vellinga and Van Verseveld of the Free University Amsterdam forecasts rainfall more akin to tropical regions and frequent flooding to become a 'normal' feature of the weather for northwest Europe. In September 2000, due to the increased instances of localized flooding, the BBC weather team introduced a flood symbol for the first time in their television forecasts. Vellinga and Van Verseveld's study, which examined a wide range of international meteorological literature and statistics from insurance companies, claims that the connection between extreme rainfall and the greenhouse effect is now indisputable.

What all of this means to farmers is becoming more clear with each passing year. You only have to ask Australian farmers who've had year on year of drought, leavened by floods. Should the disastrous scenario of the Gulf Stream stalling early in the next century (see

A desert is born

Michael Coughlan, formerly director and co-ordinator of the climate activities programs at the World Meteorological Organization, believes that the heavy floods in southern Asia and droughts across large Central Asian regions suggest climatic change rather than natural variation. He notes: 'Our records only go back 100 years. But the rainfall and drought are in the extremes of what we have recorded over that time'.

On the eastern edge of the Qinghai-Tibet plateau, a new desert is forming over what used to be rich pasture. As dunes erupt over clumps of tattered grass, the livelihoods of millions of herders and farmers who have nowhere else to go are at stake. While overgrazing has threatened the topsoil of the region, a decade of drier, warmer weather with three subsequent years of drought has pushed it over the edge. Tibetan herders from the region have been forced to become 'guerrilla grazers' according to the Chinese media, taking their animals to distant pastures already used by others.

Dr Song Yuqin of Beijing University reported the increase of barren areas in the semi-arid lands stretching from Qinghai province through to Inner Mongolia and north of Beijing. In Mongolia a drought in 1999-2000 that left livestock devastated was assumed to be the main reason why the communists returned to power in the 2000 elections with an overwhelming majority. People were looking for desperate solutions in desperate times. ■

Sources: James Poole, 'Asian floods, drought, sign of climate shift', Reuters, 12 October 2000; 'Chinese farmers see new desert erode their way of life', *The New York Times*, 30 July 2000.

chapter 1) come to pass and leave northern Europe under snow for six months at a time, current debates about agricultural adaptation would become academic.

In Britain too the effects have been felt. The winter season of 2005-2006 plunged the country into a record breaking drought. This is a bit out of sync with recent extremes of autumn/winter rainfall and summer dry spells. In a previous year, one small grower of organic vegetables wrote to me: 'We have had some very funny weather since last October; no frosts, no real cold weather only rain and more rain. Seeds and potatoes were late going in so with the rain they went rotten in the ground. This is the first time we have had to buy new potatoes in 16 years. Onions like golf balls in size, leeks that we could not get out of the ground, no marrows, no tomatoes. Beans seem to be going well, plenty of flowers, but no bees to pollinate them, so we shall have to wait and see.'

While scientific models of the changing climate's impacts on agriculture attempt to grapple with the quantity and complexities of the factors involved, ordinary people who take pride in growing things have taken note what the weather is throwing at them and they're waiting to see what happens next.

1 Jan McGirk, 'The floating garden helps farmers beat the floods', *The Independent*, 3 January 2006. **2** Devinder Sharma, 'The fight for food', *New Internationalist* No 319, December 1999. **3** IPCC, *Climate Change 2001: Impacts, Adaptation, and Vulnerability*, available at www.ipcc.ch **4** The Hadley Center for Climate Prediction and Research, 'Climate Change and its impacts: Stabilization of CO2 in the atmosphere', The Meteorological Office, October 1999. **5** Ross Gelbspan, *The Heat is On* (Perseus Books, 1998). **6** Patricia Reaney, 'Climate change raises risk of hunger', Reuters, 5 September 2005. **7** Cynthia Rosenzweig and Daniel Hillel, 'Potential Impacts of Climate Change on Agriculture and Food Supply', 1995, www.gcrio.org/CONSEQUENCES/summer95/agriculture.html **8** Peter Bunyard, 'A Hungrier World', *The Ecologist*, March/April 1999. **9** Vandana Shiva, 'Cry foul, cry freedom', *New Internationalist* No 206, April 1990. **10** *Colors*, October/November 2000. **11** Kennedy Graham, 'Consumption: Introduction', *The Planetary Interest* ed. Kennedy Graham (UCL Press, 1999). **12** Nick Sundt, 'Agriculture and Climate Change: A hard row to hoe', July 2000, www.globalchange.org/featall/2000winter2.htm

5 Impacts: wildlife and forests

The first extinction caused by climate change... coral reefs in crisis... the basis of marine life in upheaval... the threats to Polar wildlife... would animals and plants be able to adapt?

IT'S DELICATE, TINY, more orange than golden, and presumed extinct. The golden toad of Costa Rica holds the dubious distinction of being the first extinction that can be laid at the door of climate change.

A former denizen of the misty, humid cloud forests near Monteverde in Costa Rica, its demise has been blamed on rising temperatures. With the growing heat, the air has been getting less humid in the cloud forests since the mid-1970s. Clouds in the Costa Rican mountains now form at much higher altitudes, so mists are also less common. All this spelt disaster for the moist, lustrous, breathing skin of the golden toad. For many of the other frogs, toads and salamanders of the cloud forests, the changes are believed to have triggered disease outbreaks such as fungal infections that affect the skins of these amphibians. Researchers studying a 7,400-acre area (30 square kilometers) in the forest say that 20 out of 50 species of frogs and toads have disappeared. Forest lizards that are not susceptible to such fungal infections have also disappeared and climate change rather than a specific infection is viewed as a likely culprit.

It seems highly likely that there are other extinctions that have originated from climate change, but which have not been recorded. An escalating pace of change is putting additional stress on species that may be feeling the brunt of other environmental changes caused by humans, such as deforestation and water pollution. The case of the golden toad shows some of the susceptibilities that other species might also share. The toad had a limited habitat and was adapted to a set of very well-defined climatic conditions. It belonged to a family of creatures that are highly sensitive to rising temperatures and ultraviolet radiation. Frogs and toads worldwide are showing freakish deformities and skin abnormalities (including albinism), with ozone depletion and chemical pollutants playing a role alongside climate change. In January 2006, an international study concluded that the rise in temperatures in tropical highland regions had led to ideal conditions for the spread of a fungus that was wiping out amphibian populations. One third of some 1,856 amphibian species are now classified as threatened. Alan Pounds, the lead author of the study, stated: 'Disease is the bullet killing frogs, but climate change is pulling the trigger.'[1]

Climate change damages wildlife in other ways as well – by causing ripples in the food chain when certain species begin to lose ground, migrate or decline in numbers; by altering the physical environment; and by upsetting predator-prey balances, which usually result in explosions of the populations of animals that are pests to humans such as rodents and mosquitoes. Where animals are living in already fragile ecosystems, even slight changes can have far-reaching effects. But whereas some species' lives depend upon a fragile balance, others are hardier and can withstand a wider range of disturbances – cockroaches after all are said to have been with us since the time of the dinosaurs and will probably outlive human beings. Nevertheless, climate change marching hand-in-hand with loss of habitat to human settlements and the numerous ecological blunders of humankind will put considerable stress on a wide range of wildlife – often to the detriment of us human beings.

Corals lose their colors

The world's coral reefs are a case in point. They lure legions of snorkeling and scuba-diving tourists each year. The vibrant displays of these gemlike creatures are the stuff of *National Geographic* spreads. But corals don't like heat, living in waters ranging from 64.4° F to

86° F (18° C to 30° C). With temperature changes of just 1.8°-3.6° F (1-2° C) above the maximum to which they are used, they become stressed and start to expel microscopic symbiotic algae known as *zooxanthellae*, thus ending a marriage which provides them with essential nutrients and their vivid colors. Their white limestone skeleton becomes exposed and death can result if this 'bleaching' effect is sufficiently severe. In the record-breaking heat of 1998, bleaching affected most of the tropical coral reef systems – including those in Australia, the Indian Ocean, the Florida Keys, the Caribbean, the Red Sea and the Bahamas – leaving thousands of square miles of graveyard coral. In 2005, Caribbean waters recorded the longest period of elevated temperatures since monitoring began. The result was coral bleaching stretching all the way from Colombia to the Florida Keys. Further bleaching was prevented only by other record-shattering events – the highest number of hurricanes in the region that year.

Such mass events of bleaching have only been reported since 1979 and none of the indigenous communities that have subsisted alongside coral reefs for thousands of years have a name for the phenomenon in their languages, suggesting that it is fairly recent and not a part of some ill-understood natural cycle. So what apart from the lives of the corals and the appreciation of their beauty hangs in the balance here? For countries associated with coral reefs, the livelihoods of entire communities. Australia's Great Barrier Reef draws in $1.5 billion in tourist revenue, Florida's reefs $2.5 billion and Caribbean reefs $140 billion. They also form some of the most species diverse ecosystems in the world and are nurseries for fish – around 25 per cent of the fish catch in the Global South comes from coral reef fisheries, providing a vital source of protein. Seeing as fish stocks worldwide are already in crisis due to overfishing, any decline in habitat would have a considerable knock-on effect.[2]

Source: *The Ecologist*, Vol. 29, No. 2, March/April 1999; updated using WWF and Greenpeace sources, 2006

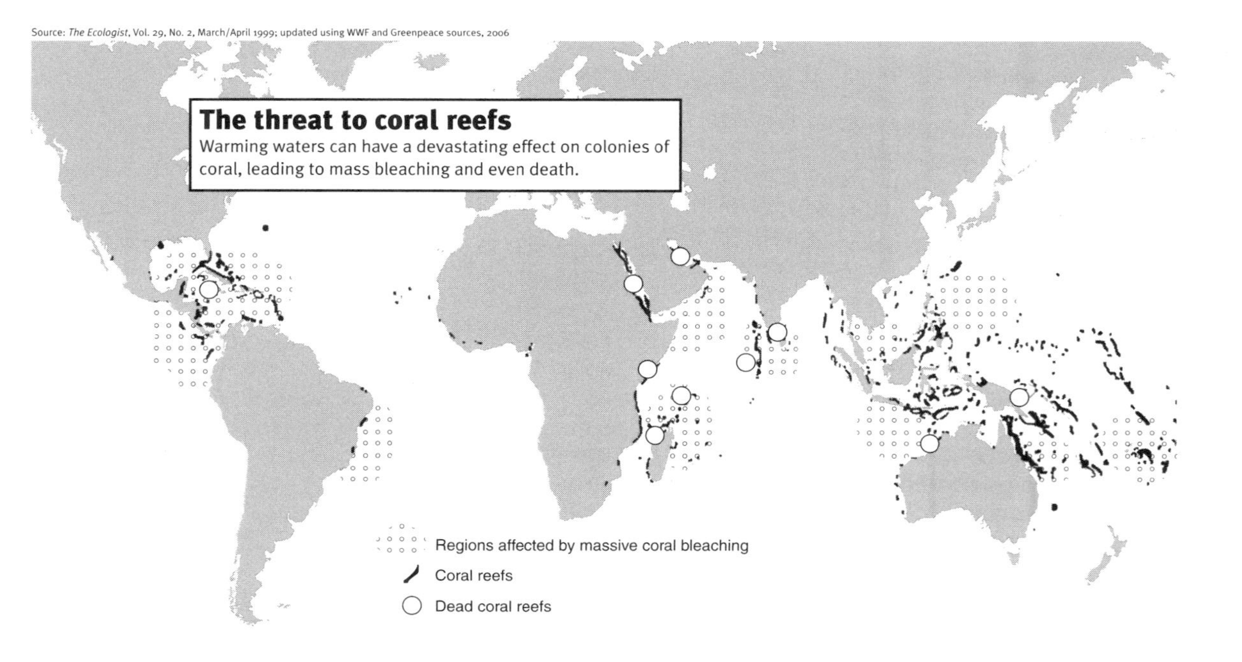

The reefs also have a role to play in defending low-lying coastal lands from storm surges and it is indeed ironic that just when the world's weather is growing increasingly stormier, this line of defense is being weakened. Not all of the damage comes from warmer temperatures – reckless human activity features in the picture as well. Rainforest destruction increases sediment run-off that can smother and kill corals. In Sri Lanka parts of the coral reef that ringed the country's coastline have been destroyed for cement production purposes. In 1986, with the monsoon in full fury, storm surges tore up beaches, houses and railway lines along the coast. The damage was more pronounced where the reef had been broken up.[3]

Whereas coral recovers in time from bleaching incidents, in extreme circumstances these can result in coral death. But with ever-increasing episodes of bleaching, recovery may become increasingly difficult and the composition of the reefs could change utterly. While corals of the genus Porites are better able to resist higher temperatures and recover fully after bleaching, others of the genus Acropora are much more sensitive. Up to 95 per cent of them may bleach and subsequently die in the months that follow. They are clearly showing little sign of being able to adapt at a fast enough rate to rising temperatures. If coral reefs reduce in size they may have their very own contribution to make to climate change, for they are yet another sink for carbon dioxide, binding it to form limestone.

We may not have long to wait. In 1999 a Greenpeace study reported the impending devastation of the world's coral reefs caused by climate change by 2030. To those who might have put this down to over-zealous environmental activism on Greenpeace's part, the IPCC findings of 2001 would come as a disappointment. They predicted the same fate for coral reefs by 2030-2050.

Fish feel the heat

Also feeling the heat are species of North Pacific salmon whose populations crashed in 1997 and 1998 as ocean temperatures hit abnormal highs – up to 10.8° Fahrenheit (6° Celsius) warmer than normal in one instance. Pacific salmon have already begun moving further north towards the Bering Sea to escape warming temperatures. In the North Sea, cod – the favorite of Britain's fish and chip establishments – are fast disappearing. Overfishing plays a major part in the North Sea cod's decline with fish often being caught before they are sexually mature and have had time to spawn, but the fish's sensitivity to rising temperatures is also indicated. A UK Ministry of Agriculture (now DEFRA) report established this link, saying the warmest temperatures in the 30 years the fish have been studied have resulted in dwindling stocks. The temperature link was demonstrated by the unusually cold spring of 1996 that caused the fish to spawn in great numbers, resulting in a bumper catch in 1998. What fishers in the region hope for more than ever is a repeat of such cold conditions, but the weather since has shown few signs of obliging.[4]

Movers and stayers

One of the most predictable impacts of climate change on wildlife has been shifts in the distribution of species, causing interlopers from what were previously warmer climes to move outwards, away from the tropics and upwards to higher elevations than before. A study tracking 14 species of European butterflies found that nine of them ranged further north by more than 120 miles (200 kilometers) during the warmer years of the late 1990s.[5] Thirty of Britain's butterfly species risk extinction. Numerous sub-tropical insects have found themselves transported in ships to Europe and are beginning to establish themselves where once they would have died of the winter cold. Muscling in on other species' territory has been reported in many marine organisms.

Apart from the phenomenon of new species moving into the happy hunting grounds of older residents and the resultant problems, there are additional difficulties raised by the shifting of temperature zones. While one may marvel at reports of Alpine plants climbing ever higher to stay within the temperature bands they prefer (though this inevitably means that the area of habitat that is suitable to them will keep diminishing in size), there are other more unfortunate species of plants and animals that are finding it difficult to keep pace with the rate of climate change. Sometimes it's because the other species that they depend upon for food or shelter have not dispersed to more suitable areas. The leanest times are for those animals that have nowhere else to go, having already reached the limits of their range.

This is the sad fate that seems to have befallen the Adélie penguin. Nesting in colonies on the shingle beaches of the Antarctic, the birds are adapted to living in what for most other creatures would be a supremely hostile environment. But warmer temperatures are leading to more snow due to increased levels of water vapor in the air. This in turn means that in the spring when eggs are laid, the snow takes longer to melt away. When the snow finally melts it can leave eggs lying in cold puddles of water or even drown newly hatched chicks. When this happens over consecutive years it can

The changing of the seasons

* Spring now arrives three weeks earlier in the US.
* In Britain 20 bird species have been found to nest on average nine days earlier than they previously did.
* In southern England, the Marsham family has kept records of the 'indications of Spring' since 1736. Thanks to their efforts it is known that the four earliest dates for oak trees to come into leaf all occurred in the 1990s.
* Tree and animal species are migrating northwards in Canada in response to warmer temperatures. The red fox has advanced 600 miles (965 kilometers) north, snapping at the heels of the Arctic fox.
* In the past 30 years, the arrival of autumn has been delayed in Britain by two days every decade. Spring's advance has been by six days every decade. ■

lead to entire colonies being deserted. Some move on to form another colony perhaps a bit further south, but there aren't many places that are suitable.

Another problem relates to a reduction in the amount of winter ice that forms on the waters where they feed. Under the ice, on the surface of the water sit algae, the first step in the Antarctic food chain. With less ice, there are fewer algae. Algae are eaten by krill that in turn form food for the penguins. So with amounts of krill also reduced the penguins begin to starve. The result has been a population decrease of a third of the total numbers in many colonies and the total disappearance of some.

When there isn't enough time to hunt

At the other end of the world a giant held in awe for its hunting prowess is being forced to go on a starvation diet. The Canadian Wildlife Service has been measuring the increasing skinniness of Arctic polar bears. Sedating the animals with tranquilizer darts, the researchers have been measuring body fat levels. The bears' major hunting season is in winter and early spring when they range across the Arctic sea-ice to gorge on seals and other marine animals, building up a layer of fat upon which they draw during the lean times when the ice has retreated. They are used to going without a feed for months at a time having hunted well during winter. But with the big freeze happening ever later in the season, the spring thaw returning earlier and the ice breaking up faster than usual, their hunting time on the sea-ice has been shortened. Meanwhile the area covered by the Arctic sea ice is also shrinking dramatically. The greatest decline was measured in 2005 – 500,000 square miles (1.3 million square kilometers) or roughly the size of Peru. This shortfall of food is not just putting adult lives at risk – the next generation is particularly vulnerable. Increasingly researchers are finding that females are unable to successfully raise the pair of cubs they usually have in a litter, during the years when they are reliant on their mother's hunting skills. By the middle of the lean season, it becomes evident that maybe one or both cubs may not make it through to the next hunting season. Birth rates are declining. From Russia come alarming first time reports of cannibalism among starving polar bears. Arctic communities are increasingly coming face to face with marauding bears in the summer months as they raid trash dumps and even homes for food. Children from these communities are taken to school under guard as they could become targets for the hungry bears.

The pack ice of the Arctic is home to several communities of animals from microscopic zooplankton to large blubbery walrus and often what isn't good for the polar bears isn't good for others as well. As the sea-ice shrinks, it spells disaster for various kinds of seal and walrus species. Walrus need to be in shallower waters to be able to dive to the bottom and feed. Upon their re-emergence the ice needs to be thin enough for them to be able to break through but thick enough to support their 2,400-pound (1,000 kilograms) weight. Seals use the ice as a convenient fishing platform, also preferring to scour the depths of shallower waters for their food. In the Alaskan Arctic, ringed seals use land-fast ice and stable sea-ice for birthing. They dig lairs in the snow that has fallen on the ice in order to protect their pups that are born in late March and early April. The pups lie here for several weeks safe from the howling Arctic winter. But should the ice break up early or rain fall to collapse the snow walls of the lairs, the chances of the pups surviving take a plunge.[6]

Seabirds gather at the interface of ice and water in the Arctic, feeding on the plentiful supply of fish. In 1997 thousands of Alaskan seabirds died as a result of unusually warm surface waters. Their fish prey, stressed by the warmer temperatures, retreated down into deeper waters, beyond the point to which the birds could dive. The birds starved to death.[6] This localized instance is symptomatic of thousands of others poised to occur or already occurring.

> ## Water birds
> The Arctic Tundra, a breeding ground for millions of water birds, could start giving way to northward moving forests as a result of warmer temperatures. More than half of all water bird populations could disappear as a result.
>
> The World Wildlife Fund/World Wide Fund for Nature (WWF) estimates that a 40-57 per cent loss of tundra would mean that 4-5 million geese and about 7.5 million Calidrid Waders would be without a habitat by the end of the century or earlier.
>
> The worst affected will be birds which are already endangered – the Red Breasted Goose, the Tundra Bean Goose, the Spoon-Billed Sandpiper, the Emperor Goose and the Greenland White Fronted Goose. ■
>
> Source: 'Climate change threatens rare Arctic Water Birds', WWF, 3 April 2000.

The seas turn to acid

Most disturbing is the extent of damage that seems poised to occur in the marine food chain. It all begins with trillions of microscopic plankton that underpin all marine life. The creatures exist in both plant (phytoplankton) and animal (zooplankton) forms. In January 2006 came reports that warmer surface temperatures were disrupting the updrifts of nutrients from the ocean floors upon which phytoplankton rely. This is alarming news on two fronts. Not only do phytoplantkton form the base of the marine food chain, but these tiny plants also remove huge quantities of carbon dioxide from the atmosphere which ends up getting dissolved in the oceans or sinking as organic waste to the seabed. So any reduction in phytoplankton would not only result in dwindling numbers of the species that depend on them, but also provide a positive feedback to a warming world.

In the North Sea, warmer temperatures have led to widespread plankton death and the migration of native species of zooplankton to cooler waters. Scientists are talking in terms of a collapse of its ecosystem. Sand eels and salmon that feed on plankton have reduced dramatically. Sea birds such as kittiwakes, guillemots, puffins and razorbills are facing disastrous breeding seasons as their favorite food – sand eels – is in short supply.

In a further twist, the US National Oceanic and Atmospheric Administration reported the results of a 15-year-long study in 2004, which concluded that the world's oceans had absorbed half of the carbon dioxide emitted by human activities in the past two centuries – at a price. They had been gradually turning more acidic. This acidification threatens all life – coral reefs, shellfish, plankton and everything that depends upon them. When scientists took snails from the less polluted waters of the Pacific near the Arctic Circle and put them in seawater with the kinds of carbon dioxide levels found elsewhere, their shells began to dissolve.

No place to call home

Loss of habitat on land is also a very real threat as droughts could affect wetland dwellers and rising seas can wipe out coastal colonies. Writing in *The Ecologist* magazine, Simon Retallack outlined the dangers: 'Sea-level rise, combined in some cases with developmental pressures [if climate change is not mitigated], will result in "about 40-50 per cent of the world's coastal wetlands being lost" by the 2080s, according to the Hadley Center – a staggering loss. Under threat are the vast tracts of tidal mudflats, salt marshes and sand dunes of the Netherlands, Germany and Denmark which are the feeding and recuperating grounds for many migrating birds... Also threatened by rising sea-levels are the wetlands of the Mediterranean, the deltas of the Nile in Egypt, the Camargue in France, the Po in Italy, the Ebro in Spain which is lived in or visited by more than 300 species of birds, and over 13,000 hectares [30,000 acres] of English shoreline... Climate change is also predicted to lead to the disappearance of the mangrove forests of the West African coast, East Asia, Australia and Papua New Guinea, which act as breeding and feeding grounds for many fish and other marine and bird species.'[7]

What climate history tells us in relation to wildlife is that gradual change is answered by corresponding adaptation – grazing herds move along the line of the shifting vegetation taking the carnivores in pursuit. Catastrophic change alters the distribution of species completely. Eventually when an ecological balance is struck again centuries later, it is a radically different one from that which went before. The phase of change that humankind appears to have engineered may not be catastrophic yet, but the danger is that it is far too rapid for adaptation to take place successfully both at a biological level as well as in terms of migration. A World Wide Fund for Nature/World Wildlife Fund (WWF) report suggests that the required migration rates for plant species due to climate change would be ten times greater than those recorded during the last Ice Age.[8]

Shrinking habitats

35 per cent of the world's existing terrestrial habitats could be destroyed by the end of this century. There can be no guarantees that they would be replaced by the formation of new habitats of similar ecological diversity, especially when one takes into account human population pressure. Extinctions would be a certainty.

Rare species or those living in unique, isolated habitats would be the first to vanish. Possible future extinctions due to climate change could include Ethiopia's Gelada Baboon, Australia's Mountain Pygmy Possum, the Monarch Butterfly at its Mexican wintering grounds and the Spoon-Billed Sandpiper at its breeding sites in Arctic Russia.

The loss of species could be as high as 20 per cent in the most vulnerable Arctic and mountainous habitats. The regions under greatest threat include parts of eastern Siberia, northern Alaska, Canadian boreal/taiga ecosystems, the southern Canadian Arctic islands, northern Scandinavia, western Greenland, eastern Argentina, Lesotho, the Tibetan plateau and southeast Australia.

The greatest loss of habitat would occur in the upper northern latitudes where the fastest rates of warming are being recorded. Up to 70 per cent of habitat could be lost in the higher latitudes of Canada, Russia and Scandinavia. Slated for a 45 per cent or greater loss are Russia, Canada, Kyrgyzstan, Norway, Sweden, Finland, Latvia, Uruguay, Bhutan and Mongolia. ■

Source: 'Global Warming and Terrestrial Biodiversity Decline', WWF, August 2000.

Fenced in

In many parts of the world, species higher up in the food chain are corralled into enclaves and reserves, surrounded by humans, often competing for the same resources. A surreal incident in March 2000 where 8 monkeys died and 10 people were injured in drought-stricken northern Kenya after a two-hour clash over a tanker of water illustrates how the fundamentals of life are the same across species. Two years later, in Australia, emus (a protected species) driven by drought were causing havoc by crashing through fences to attack farm crops. Should the climate necessitate mass migration, many animal species would have nowhere to go. As the home turf for species gets smaller their diversity gets threatened.

Shifting territories would be even more difficult for trees that rely on their seeds being carried by the wind, birds or rodents. Fossil records have been studied to reveal the speed at which plant species have migrated during changing climatic conditions in the past. They indicate that the slower species of the plant world are capable of moving about 120 feet (0.04 kilometers) a year, while the swiftest specimens can achieve speeds of 1.2 miles or 2 kilometers a year. However, just taking into account changes in temperature alone, plant species in many parts of the world would need to be able to migrate at annual rates of between 0.9-3.4 miles (1.5-5.5 kilometers). Many would not be able to stay in the race.

Smoke signals from the world's forests

It's not as if the world's forests have it easy at present. Nearly half of the original forest cover of the world is gone and trees in many regions are weakened by air pollution. While increased amounts of carbon dioxide in the air might give a boost to photosynthesis, it would also mean that trees would need better nutrients from the soil. If warmer air temperatures are added into the mix, then more moisture would be required, too. In

Vegetation dieback

When the area that was previously under vegetation has less than 10 per cent of its vegetative cover remaining it is said to be suffering vegetation dieback. The graph below shows the predicted increase in vegetation dieback as a result of increases in carbon dioxide concentrations.

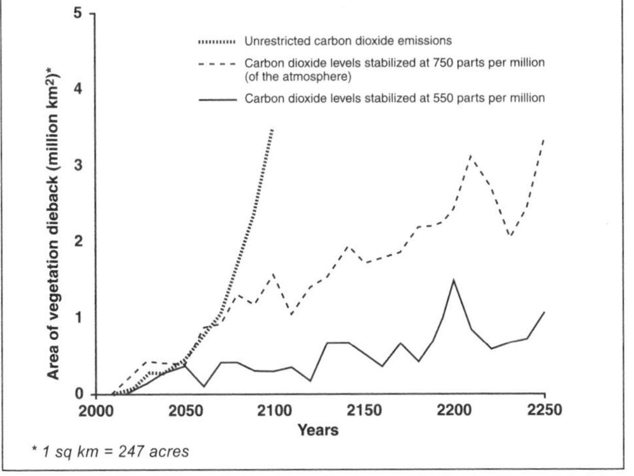

Source: The Hadley Center for Climate Prediction and Research, UK Meteorological Office.

regions with poor soils and low rainfall there could be major diebacks and loss of diversity rather than growth spurts. Warmer conditions would also favor diseases and pests. In Alaska, some 50 million acres (20 million hectares) of forest have been colonized by spruce bark beetles in an unprecedented attack, the fruit of several warm years in succession. Over 25 million trees have died as a result.[9]

Among forests identified by WWF as being particularly at risk are the boreal forests stretching through the Arctic and sub-Arctic regions of Siberia, Scandinavia, Canada and Alaska. These are at the limits of their range and changing temperatures would mean a slow decline. Animal species like the woodland caribou, the wolverine and wood buffalo would get squeezed into a shrinking habitat. The northern boreal forests have swampy soils which are a primeval reservoir of vegetation that has lain in the ground for centuries, a gigantic store of carbon that could begin to leak were the forests to be disturbed and the soils to start drying out. In western Siberia the world's largest frozen peat bog – the size of France and Germany combined – has begun to melt. It has the potential of unleashing billions of tons of methane. Sergei Kirpotin, a botanist at Tomsk State University in Russia, described it as an 'ecological landslide that is probably irreversible and is undoubtedly connected to climatic warming.' In eastern Siberia thousands of lakes have disappeared over the last 30 years, also due to climate change.[10]

Tropical forests that already come in for the chop and see the fastest rate of extinctions of species could also suffer from a combination of intense heat and lack of rainfall. They would be particularly susceptible to fire. Take the tinder-dry forests of Indonesia where fires have been breaking out every summer since 1997. Between mid-1997 and early 1998 ten million hectares of pristine rainforest went up in smoke. These fires didn't begin with dry lightning strikes: they were started by people clearing land for farming. But the fact that the fires spun out of control can only be blamed on the unusually dry conditions. The fires created a billowing expanse of smog that caused ship collisions, traffic accidents and even an airplane crash. Neighboring Malaysia lost billions of dollars with tourists canceling holidays and airports closing.[11] This single conflagration's contribution to carbon dioxide emissions was equal to all the fossil fuels burned in Europe in a year.[9] In 2005, a drought of the kind that only happens once a century, struck the Amazon region, which is already being heavily logged.

Also at risk are coastal mangroves that offer protection from storm surges to countries like Indonesia, Brazil and Bangladesh. Lying in the pathways to the sea, drawing their sustenance from submerged soils, they are fighting against rising sea levels which threaten to wash away the buildup of sediments these trees rely on.

Similarly at the edge of their limits are high mountain forests that skirt the tree line – as tree species from the lower reaches of the mountains move upwards these forests would undergo a complete change in character.

In the summer of 2000 came reports that Alaska's white spruces, the most widespread species in North America, were sending out a distress signal, by leaking carbon dioxide into the air. The trees had stopped growing so fast as a result of climatic changes that they were releasing carbon instead of absorbing it.

In 2006 came more alarming news. It had been believed that the world's forests were responsible for sinking a full quarter of humankind's emissions of

Tree trauma

The earth's forests moderate the climate, anchor soils, influence the water cycle and provide a rich habitat for myriad plants and animals.

- Half the world's original forest cover of some 7.5 billion acres (3 billion hectares) has been destroyed in the last 40 years; only 20% of what remains is undisturbed by human activities.
- More than 90% of forest loss is in the tropics; about 34.5 million acres (14 million hectares) of tropical forest are hacked down each year, two-thirds of that due to farmers clearing land.
- More than 90% of forests in the Mediterranean have been cut while from 1995 to 1997 more than 14.8m acres (60,000 sq km) of forest cover in Brazil was destroyed – an area twice the size of Belgium. ■

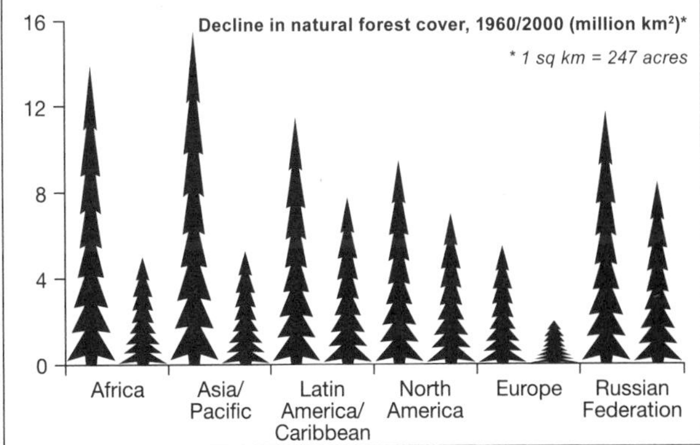

Source: Living Planet Report 1999, WWF International

carbon dioxide. Now there was disturbing new evidence that plants were actually responsible for emitting between 10 to 30 per cent of global methane.[12] The complexity of natural systems was once again commanding the respect we so patently fail to show it.

1 Steve Connor, 'Climate change is killing off amphibians', *The Independent*, 12 January 2006. **2** Ove Hoegh-Guldberg, *Climate change, coral bleaching and the future of the world's coral reefs* (Greenpeace International, 1999; 2001). **3** Peter Bunyard, *The Breakdown of Climate* (Floris Books, 1999). **4** Ian Herbert, 'Cod disappearing from North Sea as it warms', in *The Independent*, 12 May 2000. **5** Julian Pettifer, 'Edging towards meltdown', *BBC Wildlife*, March 2000. **6** Margie Ann Gibson and Sallie B Schullinger, *Answers From the Ice Edge: the consequences of climate change on life in the Bering and Chukchi seas* (Greenpeace/ Arctic Network, June 1998). **7** Simon Retallack, 'Wildlife in danger', *The Ecologist*, March/April 1999. **8** 'Global Warming and Terrestrial Biodiversity Decline', WWF, August 2000. **9** 'Nature's bottom line: climate protection and the carbon logic', Greenpeace, May 1999. **10** Fred Pearce, 'Climate warning as Siberia melts', *New Scientist*, 11 August 2005. **11** Nicola Baird, 'Breathtaking', *New Internationalist* No 319 December 1999. **12** Zeeya Merali, 'The lungs of the planet are belching methane', *New Scientist*, 12 January 2006.

6 The politics of climate change

The origins of the climate debate... political doublespeak and skeptics sponsored by fossil fuel industries attempt to sabotage negotiations... scientists issue assessments... US provides no answers... the convoluted road to emissions reductions and the loopholes in current agreements.

THERE IS A common human reaction to the possibility of terminal disease – 'It couldn't happen to me'. It exists even when there is a rational understanding of the degree of risk and one has known other people who have suffered. Sometimes it persists even after diagnosis. Denial is a common enough reaction to unpleasant truths.

Denial also features heavily in reactions to the projected catastrophes of climate change. Perhaps it's inertia, perhaps fear of the radical changes that would be necessary. Even though rationally the message may have got through that action on emissions reduction cannot wait and the issue is a burning one, we are often content to let it smolder. In fairness the enormity of the problem can be baffling, but there remains a niggling feeling that if enough people would actively voice their concern they would possibly achieve more than that achieved by two decades of political wrangling.

Ostrich arena

For it is in the political arena – where the ostrich tendency of sticking one's head in the sand to avoid doing anything seems to predominate – that the battle for the world's weather is being fought. Typical of political skirmishes, this one is dirty, shortsighted in the extreme and miles from any real breakthrough.

Scientific concern over the possible greenhouse effect of certain gases in our environment has been around since the 1970s, with scientific speculation turning slowly into scientific consensus as actual increases in carbon dioxide concentrations began to be measured. In 1979 the World Meteorological Organization brought scientists together for the First World Climate Conference in Geneva, Switzerland, which issued a call to governments to 'prevent and prepare for the negative impacts of human induced climate change'. Six years later another conference in Villach, Austria, led to the inclusion of all greenhouse gases in assessments of global warming. When this was done the doubling of carbon dioxide equivalents was estimated as early as 2030. Calls for international cooperation to head off a climate crisis gained increased urgency.

On 23 June 1988 James Hansen, a climate scientist with NASA's Goddard Institute, warned a Washington meeting that the world was getting warmer due to the build-up of greenhouse gases, and an increased tendency for droughts and floods was to be expected. He made his speech on the hottest day of that year in the US, at a time when the Midwest was in one of its worst droughts. It caught the attention of the sweltering policy-makers and population at large. Hansen's speech was a spur to the setting up by the UN of the Intergovernmental Panel on Climate Change (IPCC), a body with input from over 2,000 scientists from around the world. In their First Assessment Report in 1990 they stated that while they were unable to say that they had found the human impact on the climate as yet, it was clear that the increase of greenhouse gases would have an impact. By the time of their Second Assessment Report they would confirm 'a discernible human effect on global climate'. But right from the beginning the political battle lines were drawn, with the trillion-dollar-a-year oil and coal industries[1], the industrialized nations who were the big energy consumers and the fossil fuel producing countries realizing the threat any change in the status quo might bring. There was also another kind of politics underpinning why Hansen's speech had been taken seriously.

As physicist and activist Vandana Shiva, writing less than two years after the event, so acerbically put it: 'Thermometers registering a few degrees more in the US suddenly turned climate change into a 'global' issue. The entire scientific community was immediately mobilized.

'Contrast this with three years earlier when thousands of famine victims in Ethiopia and Sudan weren't enough to move governments in the North to respond to desertification and drought as global environmental emergencies. True, they sent food aid, but the climate problem remained a local difficulty. These deaths, after all, took place in Africa – they were still "out there".'[2]

Shiva had put her finger on a bias that persists to date – the most-detailed impact assessments are still available for industrialized countries, despite the knowledge that poorer nations in tropical and semitropical regions will suffer most. The input of these nations in the climate debate is also much more likely to be talked down.

With each passing year developments in climate change knowledge came under attack from industry-sponsored skeptics and self-interested politicians, raising a cloud of doubt over every new finding through propaganda and scare-mongering. Over the years as scientific consensus has overwhelmingly come to support the reality of climate change these voices have shouted ever louder to portray the issue as something scientists cannot agree upon and about which there is reasonable doubt. When some of them grudgingly accepted the phenomenon's reality, they then claimed only beneficial effects from global warming or drummed up scares about how emissions reductions would bring down economies.

Since the late 1970s climatologists have been attempting to work out the prospects for global warming and politicians had picked up on the issue – former Soviet leader Mikhail Gorbachev, as leader of a country whose heavy industry was notoriously energy inefficient, often spoke of the perils of a warming world. Even Britain's then Prime Minister, the right-wing Margaret

Thatcher – who had welcomed the Falklands/Malvinas War in May 1982 by saying 'It's exciting to have a real crisis on your hands, when you have spent half your political life dealing with humdrum things like the environment' – was, as the decade drew to a close, commenting increasingly on the threat to a common future posed by climate change. In November 1989 she said, with a remarkable flash of prescience, 'No issue will be more contentious than the need to control emissions of carbon dioxide... We can't just do nothing... Each country has to contribute, and those countries who are industrialized must contribute more than those who are not.'

But Thatcher's government was not about to match her words with deeds. That same month at a global warming conference in the Netherlands, Britain was a key opponent of a Dutch proposal to stabilize and reduce carbon dioxide emissions. This was despite a stated commitment by nations (albeit not legally binding) at a conference in Toronto the previous year to reduce their greenhouse gas emissions. Such duplicity was to become increasingly evident in the 1990s.

Battle commences

In May 1990 a hundred scientists of the IPCC gathered in a country hotel in the UK to sort out the wording of the final draft of their first Scientific Assessment Report. If they had hoped for seclusion, it was not to be – a media circus surrounded the hotel. Among the observers allowed to sit in on the deliberations were 11 scientists on the payroll of the fossil fuel industries. But they weren't quite observers in the traditional sense as they were allowed to make suggestions regarding the wording of the document.

Central to this first report was the link between carbon dioxide emissions and increases in the planet's temperature in the future. The IPCC at the time warned that the earth could warm by as much as 8.1° F (4.5° C) with a doubling of carbon dioxide levels by around 2050. They would later reduce the upper estimate by 1.8° F (1° C) ironically enough due to the masking effect of projected increases of pollutants in the atmosphere. The message was clear. In order to stabilize atmospheric concentrations of carbon dioxide, a 50-70 per cent cut in emissions from human activities was required. To meet such a target an energy revolution would have to come about, one that would effectively sever humankind's dependence on fossil fuel energy.

As the summary of the report (which was destined for wide media attention) was being worked on, Dr Brian Flannery, a scientist on the payroll of US oil giant Exxon, decided to intervene. The group he represented was called the International Petroleum Industries' Environmental Conservation Association, an image-conscious moniker that bears parallels with other such industry alliances and skeptical groups. Even bodies that dismiss the idea of climate change outright as nonsense are savvy enough to hide under the mask of environmental concern. Flannery suggested there were uncertainties about how carbon behaved in the climate system. The response from the IPCC scientists was that such uncertainties should not affect the ultimate goal of stabilizing the amount of carbon dioxide in the atmosphere, which could only be realized by radical cuts in emissions. Now Flannery changed tack and questioned the validity of the climate models the IPCC had used, calling their range 'scientifically uncertain'. Again the IPCC scientists begged to differ. Flannery's intervention was relatively mild compared to what the climate change skeptics would produce in years to come but it was perfectly in character.

The day after Flannery's intervention Margaret Thatcher held a press conference at which she called the report 'an authoritative early-warning system'. Warming to the theme, she spoke of some of the consequences: 'There would surely be a great migration of population away from the areas of the world liable to flooding, and from areas of declining rainfall and therefore of spreading desert. Those people will be crying out not for oil wells but for water.'[3]

To the fossil fuel industries, action on the emissions reductions front would come as a body-blow and they began organizing their armies of skeptics, PR spin-merchants and sympathetic politicians (these latter often represented constituencies where a large proportion of the workforce was involved in the fossil fuel industry and/or had received generous donations from the oil and coal giants). At stake were vast empires funded by the unhindered profits of a century of oil burning. The oil companies had already tracked down reserves holding more than the amount humans had consumed throughout history. The coal industry was sitting on even larger reserves. Neither was about to let its lifeblood lie in the ground.

From now on both scientific discussion and political negotiations were open targets and when important meetings were held, along with the scientists, environmental non-governmental organizations (NGOs) and politicians would come members of what has been dubbed the 'Carbon Club'. They were invariably fronted by a handful of skeptical scientists acting as mouthpieces for seemingly innocuous sounding groups such as the Global Climate Coalition (GCC) and the Global Climate Council. The GCC's list of membership and donors read like a who's who of the major fossil fuel producers and consumers. The Global Climate Council was decidedly more secretive about its funding, but not in the least bit less vocal for all that.

The conspicuous persuaders

Such industry-funded scientists then took center-stage in any climate debate, turning in storming media performances. Against the vast array of IPCC scientists from all over the world, this motley group consisted of only a small number of very vocal nay-sayers, most of whom were not Cassandra-like prophets of doom crying in the wilderness but rather scientists whose research was not peer-reviewed or, allegedly, had been discredited. Some had not conducted any original research in years. With just five industry-friendly media companies dominating the world's news channels, the skeptics were set for a long ride.

They used several factors to their advantage. Most news desks aim to maintain some kind of 'objectivity' by giving space to opposing sides of an argument. In this case one of the 2,000-plus pro-climate change scientists would get the same amount of airtime as a representative from the handful of skeptics or an industry spokesperson. The skeptics were adept at getting the message across that there was little scientific consensus on the issue, when in fact the world's leading climatologists overwhelmingly agreed on the reality of climate change.

Industry also pulled no punches, confidently asserting that there was no danger, that any weather extremes that happened were just an expression of natural variability and had nothing to do with climate change, and that if the phenomenon did come about it would be actually beneficial. Spurious arguments would often be unnecessarily highlighted – such as the fact that as carbon dioxide levels rise in the atmosphere their ability to trap heat progressively declines. But such knowledge was already being taken into account in the IPCC's calculations. In the US the skeptics even argued that research shouldn't be funded by government departments, but should be left in the hands of the companies most affected by it – a bit like hiring a wolf to take care of sheep.

The confidence of the skeptics who made very definite pronouncements was in marked contrast to the rest of the scientific community who talked in terms of probabilities and possible outcomes and percentages. Thus in the general perception scientists appeared to be uncertain on the issue when in reality they weren't

Packaging Mitch

In the aftermath of Hurricane Mitch in 1998 came a wave of denial about its possible origins. *Sheldon Rampton* and *John Stauber* recount one skeptical voice's determination to shout louder than the rest.

It was only natural to wonder if global warming was to blame for the disaster. 'This was perhaps what is becoming a typical disaster in today's world of El Niño and global change,' observed J Brian Atwood, head of the US Agency for International Development, which coordinated relief activities. Speaking to CBS News, he called the hurricane 'a classic greenhouse effect'.

For Patrick J Michaels, people like Atwood are part of the problem. Michaels, professor of environmental science at the University of Virginia, penned an article titled 'Mitch – That Son of a Gun'. He attacked Atwood's remarks as 'White House huckstering... If there's any possible way to conflate human suffering with global warming, the Clinton administration will do so... Rumors persist that Vice President Gore has been advised to make global warming a central theme of his presidential run in 2000. Threatening hundreds of thousands with imminent drowning unless they vote for him is a crude but probably effective trick.'

Michaels' commentary was printed in the *Washington Times* and the *Journal of Commerce*. Rewritten as local editorials, it appeared in newspapers as far apart as the *Wisconsin State Journal* and the *Wyoming Tribune-Eagle*. 'Just how stupid does the Clinton administration think we are?' asked the version that appeared in the *Tribune-Eagle*.

Stupid enough, apparently, that none of these outlets bothered to check Michaels' credentials. If they had, they would have found that Michaels is part of a small but vocal minority of industry-funded climatologists who dispute the mounting evidence that suggests that global warming is a consequence of modern industrial activities, such as the burning of fossil fuels. By his own account, Michaels has received more than $165,000 in funding from fuel companies, including funding for a non-peer-reviewed journal he edits called *World Climate Change*.

The use of scientists as spokespersons for corporate interests is an example of a public-relations strategy known within the trade as 'the third party technique'. Merrill Rose, executive vice-president of the public-relations firm Porter/Novelli, sums it up succinctly: 'Put your words in someone else's mouth.' Remember the TV commercials with actors in lab coats pretending to be doctors and claiming that nine out of ten of their colleagues prefer a specific brand of aspirin? With commercials you are on your guard. But put the message in the mouth of someone like Patrick Michaels and you have a 'real scientist' speaking. The commercial interests behind the message are much better disguised. ■

From 'The junkyard dogs of science', *New Internationalist* July 1999.

and, worse, the skeptics sounded more convincing.

As if this wasn't bad enough, the political response to the IPCC's findings in no way reflected the urgency demanded by the science. In August 1990 after the fuss created by their first scientific assessment, the IPCC met again with government representatives in Sweden. Scientist and campaigner Jeremy Leggett's account of this meeting in his book *The Carbon War* is revealing: '... the political geography was already clear. The IPCC's scientific working group had professed itself "certain" that global warming lay ahead unless greenhouse-gas emissions were cut. The impacts working group had predicted a collage of expanding environmental catastrophe should the IPCC scientists' predictions turn out to be correct. But all that the policy responses working group had come up with, after 18 months of deliberation, was a toothless list of potential technologies which could help, in principle, with the limitation of greenhouse gases. This third working group was chaired by the United States.'[3]

Here it is pertinent to consider one fact: the US with 4 per cent of world population accounts for nearly a quarter of the world's greenhouse gas emissions. It follows that the deepest emissions cuts would have to be made by the US, a task most of its politicians have been loath to contemplate.

False parameters

After a series of governmental meetings, the stage was set for the UN's 1992 Rio de Janeiro Conference on Environment and Development (UNCED) at which more than 160 countries signed the Framework Convention on Climate Change before the largest media circus ever gathered in the world, numbering over 10,000 press representatives. The Framework Convention, designed to set industrialized countries down the path of emissions reductions, may have been a setback for the Carbon Club, but it was one from which they devised a strategy that was to have far-reaching repercussions. They now started pointing out the potential for future emissions increases by developing countries, thus attempting to deflect attention from the largest polluters.[3]

The trick was to arouse enough indignation on the issue so that any discussions about emissions cuts by the industrialized nations would have to be linked to parallel cuts in countries like India and China whose emissions potential is great. This idea stuck like a thorn in the side of subsequent climate change negotiations. However, it is estimated that historically up to 90 per cent of greenhouse gases in the atmosphere arising out of human agency were emitted by the 20 per cent of people who live in industrialized countries. Having fossil fuel energy to thank for their global supremacy today, it is hypocrisy of the highest order to suggest that industrializing countries are on a level playing field and must compromise their future development in order to cushion the excesses of the rich West. This is not to write a blank check for Majority World countries to pollute, but to recognize their right to develop. Many countries have already signaled their willingness to clean up industry, work on curbing emissions, install cleaner energy production – they just want it at a price they can afford.

Towards the end of 1995 the IPCC completed its Second Assessment Report which found that 'a pattern of climatic response to human activities is identifiable in the climatological record'. It put paid to the argument that natural climatic variability was to thank for the heated 1980s and 1990s.

All fired up

However, none of this would faze Republican politicians in the US who, wooed by the oil and coal lobby, have consistently taken up cudgels to defend both industry's and the ordinary American's right to pollute. In the summer of 1995 Republican Member of Congress

Robert Walker was the leading light in arguing for cuts in funding for a NASA program that aimed to monitor climate changes around the world. He succeeded in his mission, relying heavily on the findings of the Washington-based George C Marshall Institute that had confidently asserted that 'comparable temperature changes are commonplace in recent climate history'. According to environmental journalist Ross Gelbspan, this is an organization 'which conducts no original research itself and whose reports are viewed by the vast majority of scientists as political statements rather than as research contributions.'[1]

Republican Dana Rohrabacher went one better when he presided over a series of House Science Subcommittee on Energy and the Environment hearings. Upon listening to an Environmental Protection Agency official talk of the potential rise in sea level over the next century being capable of drowning up to 60 per cent of the US's coastal wetlands, Rohrabacher's sage reply was, 'I am tempted to ask what this will do to the shape of the waves and rideability [sic] of the surf.'

In a hearing whose title – 'Scientific Integrity and the Public Trust' – was no doubt not intended to be ironic, Rohrabacher threw further pearls to his audience. Speaking about the ozone scare, he said it 'turned out to be another basically the sky-is-falling from an environmental Chicken Little, a cry we've heard before when the American people were scared into the immediate removal of asbestos from their schools.' That someone chairing a meeting on scientific issues should dispute the dangers of asbestos is beyond belief. At one point in the proceedings Rohrabacher found himself unable to remember the word 'hydrocarbons' and started speaking of carbohydrates instead.[1]

A way with words

The most quoted line from the IPCC's Second Assessment Report was 'the balance of evidence suggests there is a discernible human influence on global climate'. The elegance of its phrasing belies the heated discussions that went into key words. It had started life simply as, 'the changes point towards a human influence on climate.' But this was much too direct for the interested parties and 'the changes' got converted into 'the balance of evidence' with its implicit suggestion that there was significant evidence that pointed in the other direction as well.

Again 'a human influence' was too direct and one of the scientists suggested 'an appreciable human influence' before 'discernible' became the adjective to find favor. Delegates representing the Saudi Arabian and Kuwaiti governments, and briefed by prominent skeptic Washington attorney Don Pearlman, objected frequently. At one point Mohamed Al-Sabban, an oil ministry official from Saudi Arabia and veteran of climate negotiations suggested that 'Preliminary evidence which is subject to large uncertainty points towards a human influence.' The IPCC scientists, used to proceeding by scientific consensus found such negotiation over almost every line of their text both novel and wearing. So wearing in fact that by the time it came to be published, the executive summary – the part that would be most widely consulted by the world's media – had shrunk to a quarter of its original size mainly due to the constant stalling by vested interests.[3]

Between the discussions of the Report text and its appearance in print six months later in June 1996 the skeptics went into overdrive in their attempts to discredit the IPCC's findings. There were also personal attacks on some of the IPCC scientists which were to alert the scientific community to the kind of mud-flinging the skeptics were capable of (see box: Man in the middle).

However, after the publication of the Second Assessment Report the issue of emissions reductions came to be firmly on the agenda at the Conference of Parties (COP) meetings of the Climate Change

Man in the middle

When the UN's Intergovernmental Panel on Climate Change (IPCC) was working on its Second Assessment Report in 1995, the industry-funded skeptics were unable to provide any compelling evidence to disprove the Report's findings. However, they picked their chance by rounding on Dr Benjamin Santer who had been given the task of making changes to the text so that it would match the wording of the summary more closely. The summary itself had emerged after much debate and wrangling.

In May, a month before the print version of the report appeared, Santer addressed a symposium with another IPCC scientist explaining the findings of the IPCC and the workings of the climate models they had used to a capacity crowd. At the end of his presentation, William O'Keefe, chair of the Global Climate Coalition and Don Pearlman laid in with a barrage of accusations. Santer, they said, had secretly changed the IPCC report, excising any dissenting voices and references to scientific uncertainties. Santer and his colleague's replies that one section of the report had been moved so that the document would be more accessible and that the chapter had been written not by Santer alone but by 40 scientists and reviewed by another 60 was met with outright disbelief.

If Santer was shaken by this very public attack, there was worse to come. With the publication of the Report came attacks in *The Washington Times*, *The Wall Street Journal*, and *The New York Times*. Santer was accused of politically 'cleansing the underlying scientific report' and 'a most disturbing corruption of the peer review process'. The aim of these unsubstantiated reports was not just to isolate and attack Santer, but to discredit the Report itself. Santer sent a letter to each of the Report's authors apprising them of the allegations. 'I had hoped that any controversy regarding the 1995 IPCC Report would focus on the science itself, and not on the scientists. I guess I was being naive.' Several of the IPCC scientists wrote in defense of Santer's integrity to the papers including chair Bert Bolin who in a letter to *The Wall Street Journal* stated 'No one could have been more thorough and honest.' ■

Sources: Ross Gelbspan, *The Heat is On* (Perseus Books, 1998); Jeremy Leggett, *The Carbon War* (Penguin, 1999).

Convention. Despite clouds of disinformation the issue was not going away. In the US, the fossil fuel lobby tried a variety of tactics from releasing alarming stories about outrageous energy bills if the status quo were disturbed to funding the teaching of 'how petroleum improves the quality of life' to schoolchildren.[4]

As the reality of climate change became more difficult

to deny scientifically and with the GCC's dirty tactics coming under increasing scrutiny, British Petroleum (BP) withdrew from the group in October 1996, with Shell following two years later. Both though continued to be members of other lobby groups which in turn were members of the GCC. In 1997 BP committed $1 billion to solar energy, while Shell allocated $500 million to renewable energy options. Such good news must be seen through the filter of business reality. It could be interpreted as these two giants finally catching a glimpse of a future in which fossil fuel had but a small role to play and getting in early in the race to develop viable alternative energy options. But there was no concomitant commitment to reducing the scale of their oil exploration. Environmental critics also derided the sums allocated – when taking into consideration that US businesses alone spend $500 million each year in 'greenwashing' their image. They also noted that as the global oil industry has a turnover of over $2 billion each day, the sums Shell and BP had allocated seem not quite so considerable. Meanwhile the mileage they got out of portraying themselves as 'green' corporations was considerable. One can also argue that both companies' subsequent engagement with the climate talks as 'supporters' of the Kyoto Protocol (see below) has actually been a clever tactic to water down the agreement and obstruct.

Stalling strategies

In 1997 with an important meeting of the Climate Convention looming in Kyoto, Japan, with targets and timetables for reducing emissions of greenhouse gases on its agenda, stalling moves were being set in motion by a resolution in the US Congress. Co-sponsored by Senator Robert Byrd, a Democrat from the coal state of West Virginia, and Senator Chuck Hagel, a Republican from Nebraska, a state whose farmers would stand to lose much from climatic adversity, the resolution proposed ruling out any action on climate change by the US unless developing countries also 'participate meaningfully' in taking steps to reduce their greenhouse-gas emissions.

It also ruled out action that would 'harm' the US economy. These were old bogeys that were revived at precisely the time when they would do the most damage – months before representatives of the world's governments would be gathering to attempt to thrash out a deal on emissions reductions. The resolution threw out the arguments about equity that had framed the convention in favor of naked self-interest. Byrd argued that if the US agreed to reduce emissions, industries and jobs would flow right into the countries in the Majority World if they didn't have similarly binding commitments from them. But with two-thirds of US emissions arising out of the building and transport sectors, it was unlikely these jobs were at risk of moving anywhere.

That left the one third which was emitted by industry, a sector which the Government could buffer with incentives rewarding energy efficiency.[3]

Hagel is perhaps the ultimate climate change skeptic who rejects the science behind the greenhouse effect outright, saying it is 'unproven historically – it doesn't make any sense', and who denies that temperatures worldwide have been rising, despite concrete evidence to the contrary. The bottom line is about the impact that emissions reductions could have on the engine of big business. Again and again the argument surfaces that economic growth cannot in any way be compromised by the threat of impending climate catastrophe, instilling a kind of wait and see mentality that plays right into the hands of the skeptics. Professor of Environmental Sciences University of Virginia Patrick Michaels and others of his ilk often argued that they would believe climate change is for real when they see conclusive proof – a bit like waiting to see the smoke coming out of the barrel of a gun before declaring that you've been shot.

While Byrd and Hagel could not be said to have acted from any sense of altruism, their actions possibly were not just about saving the US economy from an imagined threat. It's just possible that they might have been looking after themselves, what with Byrd having received $199,700 in political contributions from fossil fuel related industries in 1996 alone and Hagel totting up $148,000 that same year.[5] Of course the Byrd-Hagel resolution could have been defeated by right-minded politicians, but senators (including almost all Democrats) lined up to vote in favor. The suggestion that somehow developing countries would otherwise gain an 'unfair' advantage had done the trick.

Daphne Wysham, a researcher with the Washington-based Institute for Policy Studies which has prepared several critical reports on World Bank involvement in fossil fuel projects, captured the mood in the US Senate when she said, 'Climate change is like the new Communism,' adding 'I've been told by people in the Treasury Department that we cannot mention the words climate change in our language [to Congress] in our appropriations for the World Bank; if we do it will be struck from the record.'[5]

With the Byrd-Hagel resolution still fresh, the Clinton administration, which had previously made noises suggesting some commitment to the issue of climate change and whose Vice-President Al Gore had once written an urgent call to action entitled *Earth in the Balance*, seemed in a decidedly wobbly state to make any positive headway on the issue of emissions reductions. In 1992 Gore wrote: 'Minor shifts in policy, marginal adjustments in ongoing programs, moderate improvements in laws and regulations, rhetoric offered in lieu of genuine change – these are all forms of appeasement designed to satisfy the public's desire to believe that sacrifice, struggle and a wrenching transformation of society will not be necessary.'

Now it was time for the US delegation to Kyoto to play just that kind of dodging game that Gore had so eloquently condemned, with presumably his

Who pays the piper?

The close relationship between fossil fuel producers and US politics is seen as a major sticking point when it comes to making progress on enacting measures that would benefit the environment. Further, the amounts spent by the companies to lobby political representatives in the US dwarfs even their generous contributions to party coffers.

ExxonMobil			
	Lobbying	1998-2004:	$59,672,742
	Lobbying	2004:	$7,700,000
ChevronTexaco			
	Lobbying	1998-2004:	$32,803,755
	Lobbying	2004:	$5,220,000
Royal Dutch Shell			
	Lobbying	1998-2004:	$26,608,088
	Lobbying	2004:	$1,020,320
BPAmoco			
	Lobbying	1998-2004:	$26,793,984
	Lobbying	2004:	$2,281,000
Occidental Petroleum			
	Lobbying	1998-2004:	$13,575,400
	Lobbying	2004:	$2,027,964
Marathon Oil Corp.			
	Lobbying	1998-2004:	$29,190,000
	Lobbying	2004:	$1,890,000
All Oil and Gas Companies			
	Reported Lobbying	1998-2004:	$343,896,623
	Reported Lobbying	2004:	$45,459,024
American Petroleum Institute			
	Lobbying	1998-2004:	$17,395,630
	Lobbying	2004:	$2,870,000

Oil and gas industry contributions to US political parties

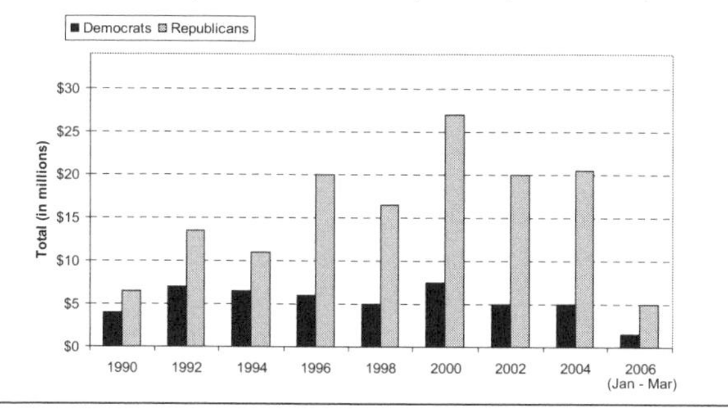

Sources: Center for Responsive Politics, www.opensecrets.org: Center for Public Integrity, www.publicintegrity.org

full knowledge and consent. (Gore would also be instrumental in pushing for market-based mechanisms, citing 'the magic of markets', which have become such a dead end as far as solutions to the climate crisis are concerned.) And they were not the only country playing it. But it appeared that the views that the US delegation were to express at the Kyoto conference were not the views of ordinary Americans who were increasingly making the connections between tornadoes, droughts, weird overnight epidemics, explosions in the insect and rodent populations and the weather. For on the eve of the Conference *The New York Times* published the results of a large opinion poll on global warming – 65 per cent of the people polled felt that the US should cut its emissions immediately regardless of the position of other countries.

Division in Kyoto

As the Kyoto conference opened, the country positions seemed more or less clear from the outset. The view

The finger of blame

Historically, industrialized countries have contributed up to 90 per cent of the greenhouse gases that are tipping the world's weather system out of joint. Today, the US, with 4 per cent of the world's population, emits nearly a quarter of all greenhouse gases.

This graph shows the per capita amount of carbon dioxide sent into the atmosphere by various countries and regions. ■

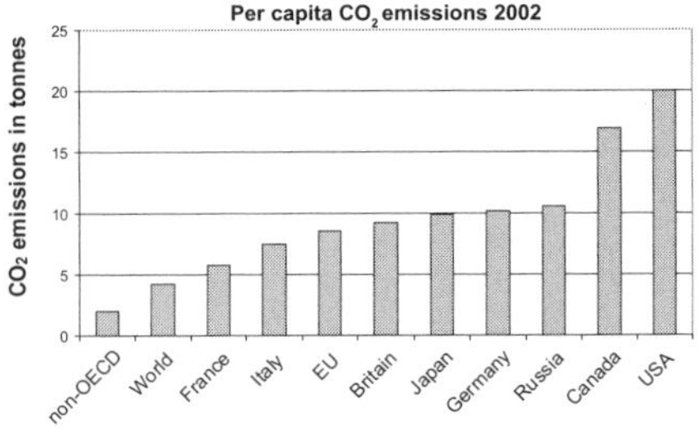

Per capita CO_2 emissions 2002

CO_2 emissions in tonnes

non-OECD, World, France, Italy, EU, Britain, Japan, Germany, Russia, Canada, USA

US ton = 2,000 lbs. 1 metric tonne = 2,240 lbs/1,000 kg.

*OECD = Organization for Economic Co-operation and Development of 30 industrialized member countries.

Source: International Energy Agency

of the OPEC (oil-producing) states, which had the most to lose, was that any emissions reduction target should be a low one. These oil-producers of the Middle East, fully briefed by the skeptics, set their faces firmly against any cuts that would threaten their livelihoods. Britain with Germany led the European call for cuts. However Iceland, with vast reserves of natural gas, and Norway with its North Sea oil reserves were notable dissenters arguing for an expansion of their emissions limits. Canada, whose per capita emissions are the second-highest, toed the US line of asking for parallel commitments from the global South; while Australia, the world's largest coal exporter, argued for an increase in its emissions. The industrializing nations were keen to defend their right to develop.

The strongest cuts in emissions were proposed by the Alliance of Small Island States, whose member countries are at risk of sinking under rising seas. They argued for cuts of 20 per cent below 1990 levels, the baseline year agreed at Rio. Ambassador HE Tuiloma Neroni Slade of Samoa spoke with passion: 'The strongest human instinct is not greed – it is not sloth, it is not complacency – it is survival... and we will not allow some to barter our homelands, our people, and our culture for short-term economic interest.' But that did not prevent some countries suggesting that it would be cheaper to relocate the populations of some of these smaller island states than do something about emissions, a suggestion decried by the Prime Minister of Tuvalu as 'utterly insensitive and irresponsible'.

Predictably much of the wrangling was duplicitous in the extreme. OPEC members such as Saudi Arabia and Kuwait and their industry counterparts stuck to their line of trying to persuade large nations like India and China that climate change was just a scare, a plot of the wealthy countries to keep them poor. On the other hand they made it clear that they themselves would accept no cuts unless similar cuts were imposed on developing nations as well.[1] China had been visited by Exxon Chief Executive Lee Raymond little over a month earlier. In his address to the World Petroleum Congress in Beijing he did a hatchet job on the idea of global warming and encouraged China to exploit its fossil fuels to the fullest in order to achieve economic progress. Back in Washington Exxon had been financing the Global Climate Coalition's campaign to push developing countries to commit to cuts at Kyoto.[3]

When the US and Canada insisted upon Majority World countries pledging reductions, Zhong Shukong – the Chinese delegation leader – replied: 'In the developed world only two people ride in a car and you want us to give up riding the bus!' He was right: at that time Los Angeles alone had more cars than the whole of China.

Commitments to change

Out of this tangle of positions and despite the efforts of the industry-fueled skeptics, a target commitment did emerge from the wealthy nations to cut their greenhouse gas emissions by 5.2 per cent below 1990 levels by 2012. In terms of effects on the world's temperature this would mean a mere 0.2° F (0.1° C) improvement in the projected warming of 2.7° F (1.5° C) over the coming 50 years according to the IPCC models of the time.[6] Australia, Iceland and Norway were the exceptions, winning increases in their emissions targets. The US committed itself to a 7 per cent cut while the EU target was 8 per cent. Even though the overall target was a far cry from the 70 per cent reductions that are needed to stabilize world climate and which would mean in effect a phasing out of fossil fuels, it was still an improvement on the business-as-usual scenario where emissions would be allowed to grow unchecked.

Kyoto was widely perceived at the time as a beginning, whatever the disappointments over low targets. But industrialized nations were loath to ratify it. President Clinton did not dare utter a squeak about ratification in a Republican-dominated Congress. His successor had

Low commitments

After the 1997 climate conference in Kyoto, the industrialized countries of the world made a commitment to cutting their greenhouse gas emissions which averaged at 5.2 per cent below 1990 levels. They were to achieve this by 2012. This was nowhere near the 50-70 per cent cuts needed, but was viewed at the time as a start.

Here are how some individual countries have fared.

Britain

- Target 12.5 per cent below 1990 levels.
- Looked on course to achieve this due to an earlier transition from coal-fired power stations to gas, but recently emissions have begun to rise again.

Canada

- Target 6 per cent below 1990 levels.
- 22.6 per cent above 1990 levels as of May 2005.

France

- Was supposed to match 1990 levels.
- Despite reliance on nuclear power keeping emissions low, will not be able to meet target without carbon trading.

Germany

- Target 21 per cent below 1990 levels.
- Looks likely to meet it. However, much of the reduction is due to the economic slump in the Eastern part of the country during the 1990s.

Italy

- Target 6.5 per cent below 1990 levels.
- In 2002 emissions were 8.8 per cent higher than in 1990.

Japan

- Target 6 per cent below 1990 levels.
- In 2002 emissions were 12.1 per cent higher than in 1990.

Russia

- Stabilize at 1990 levels.
- Economic collapse means Russia will remain well below this. Remains a fossil fuel exporter.

The US is no longer within the Kyoto Protocol so its Kyoto target of a 7 per cent reduction below 1990 levels doesn't hold. Despite impressive moves in many states and at the city level, emissions will be much higher than 1990 levels. ■

Source: Carbon Trade Watch, *Hoodwinked in the Hothouse: the G8, Climate Change and Free-Market Environmentalism*, June 2005, TNI briefing series No 2005/3.

clearly no desire to do so – quite the opposite. Oil had played a significant part in George W Bush's checkered career. He quickly made known his opposition to the modest limits set at Kyoto. His early response to the problem of global warming sounded uncomfortably close to that of the fossil fuel lobby – apparently he thought the matter needed 'more research'.

Loopholes: old habits die hard

The industrialized world's politicians have been busily studying how best to use the loopholes which they had created in the Kyoto Protocol that would allow them to carry on business as usual while making claims to doing something about the environment.

The US was instrumental in pushing through an Article in Kyoto that would allow it and other industrialized countries that may have difficulty meeting their targets to trade emissions with countries that have generous targets. Cases in question are Russia and Ukraine, which as a result of the collapse of their economies are already down on their 1990 levels of emissions. They were allowed increases of 50 and 120 per cent respectively under the Kyoto Protocol but it is highly unlikely that they would be able to use them. At this point companies in foreign countries could offer to buy some of their permissible emissions in order to carry on polluting without check. Since the emissions allowance was a theoretical increase anyway, such trading translates it into an actual increase in emissions while letting the purchaser claim they are meeting their targets. Emissions trading is flawed in principle as well, as it acknowledges the 'rights' of the worst polluters to pollute and has a built-in potential for fraud.

Another bone of contention is the possible unlimited use of carbon 'sinks' in the accounting; this could mean that countries do nothing to actually reduce fossil fuel emissions. Such sinks are primarily seen to be the planting of new forests, but the creative accounting involved is, well, creative. The use of sinks could actually encourage cutting down old-growth forests in order to gain carbon 'credits' from ecologically disastrous fast-growing monoculture plantations or genetically engineered tree species.[7] Further there's no scientific foundation to the use of such plantations as carbon sinks. Increasingly they are being planted in Majority World countries with no consultation of local people (often dispossessing them) and little regard for the local ecology. Commentators call it carbon colonialism.

Yet another loophole is the provision of emissions-reduction technologies to Southern countries in order to claim the estimated emissions cut in pollution in the home country, the so-called Clean Development Mechanism. With this, wealthier countries would enable some emissions reductions (and quite possibly lucrative contracts for the alternative energy industry riding piggyback on such 'gifts') in the Majority World and themselves gain the right to pollute more. But there is no one to police whether the emissions cuts claimed have already been delivered. Further such moves deprive countries of the global South of agency – when they commit to specific emissions reductions targets themselves after 2012 (in the second phase of the Kyoto treaty) they may find themselves saddled with inappropriate technologies and forced to pay for more expensive ones.

These loopholes benefit most those with the dirtiest hands. As climate campaigner Heidi Bachram puts it: 'The Kyoto Protocol negotiations have not only created a property rights regime for the atmosphere. The Protocol has also awarded a controlling stake to the world's worst polluters, such as the EU, by allocating credits based on historical emissions.' (She doesn't mention the US as it has pulled out of the Kyoto Protocol – see below.)[8]

In June 2000 a global coalition of environmental organizations including Greenpeace, World Wide Fund for Nature/World Wildlife Fund (WWF) and Friends of the Earth (FoE) voiced their concern that the 5 per cent reduction promised by the industrialized nations was in danger of being twisted to a 15 to 20 per cent increase, singling out the US, Canada, Japan, Australia and New

Smokescreen

The World Bank's remit is the alleviation of poverty and the promotion of sustainable development. So it's understandable that about a fifth of its lending goes towards energy generation in poor countries. But if sustainability was really that high on its agenda it wouldn't be spending 17 times more money on fossil fuel related projects than it did on renewable energy ones. Each dollar the World Bank pays out to fossil fuel projects attracts up to five to six more in private investments. World Bank supported projects in Nigeria and Chad have encouraged the export of fuel to wealthy countries. Over 80 per cent of all oil projects financed by the World Bank since 1992 are for export to the rich world. When fossil fuel projects generate power in the home country it is usually allocated to urban populations and industry, including energy-hungry industries such as aluminum production, which often move in from other countries as soon as cheap energy becomes available. The rural poor usually remain unaffected by any benefit.

Between 1994 and 2003 the Bank approved over $24.8 billion in fossil fuel projects with just $1.06 billion earmarked for renewable energy projects.

In China, the World Bank spent $1.35 billion on building coal-fired power plants in the 1990s. Seeing as coal is the fuel that emits the largest amount of carbon dioxide per unit of energy produced it would appear that the World Bank cares little for its own sustainable development policies. The US is the largest stakeholder in the World Bank, providing 18 per cent of its funds and could if it wanted, redirect its investments to China away from coal. In not doing so it can be seen as encouraging China to pollute.

The Bank is also active in carbon trading schemes and has developed around $41 billion worth of carbon transactions from which it will earn commissions. According to Brazilian activist Marcelo Calazans: 'In an absurd contradiction the World Bank facilitates these false, market-based approaches to climate change while at the same time it is promoting, on a far greater scale, the continued exploration for, and extraction and burning of fossil fuels – many of which are to ensure increased emissions of the North.' ■

Sources: Jim Vallette, Daphne Wysham and Nadia Martínez, *A Wrong Turn from Rio: The World Bank's Road to Climate Catastrophe*, a report by IPS, SEEN and TNI, 10 December 2004; Carbon Trade Watch, *Hoodwinked in the Hothouse: the G8, Climate Change and Free-Market Environmentalism*, June 2005, TNI briefing series No 2005/3; Kate Hampton, 'Smokescreen', *New Internationalist* December 1999.

Zealand/Aotearoa as the 'main culprits'. Bill Hare, the Climate Policy Director of Greenpeace International, was particularly withering: 'These governments are trying to create the impression that they are moving ahead on climate policies while in reality, in the smoke-filled back-rooms of these negotiations [in Bonn, June 2000], they are systematically attempting to shred every last bit of environmental integrity from the Kyoto Protocol.'[9]

Facades and horse-trading

Just how big a facade the Kyoto Protocol was in danger of becoming was revealed at the Sixth Conference of Parties (COP 6) meeting at The Hague in the Netherlands during November 2000. With the Gore versus Bush ding-dong battle for the White House playing out back home, the US delegation arrived intent on clinging to the loopholes which it was the Conference's purpose to close. Right from the start American negotiators fielded searching questions by the world's media with an oleaginous slickness, insisting that a deal would be made and genuine compromises offered without budging from their position that their seven-per-cent Kyoto target would only be met by accounting for 'sinking' or absorption by forests and not plowing farmland.

Frank Loy, one of the US negotiators, stated he would do nothing that would 'jeopardize the American lifestyle' and advanced the hard-line economic argument that the Majority World's interests would be best served by an expanding US economy. Although it was the US position that finally led to the collapse of the talks, Japan and Canada were two other important players who were riding on the US delegation's coat-tails. The EU group of nations was determined to preserve some semblance of environmental integrity for the Kyoto Protocol.

As deliberations wore on without any sign of a breakthrough and the conference overshot its closing deadline, British Deputy Prime Minister John Prescott received permission to undertake some last minute horse-trading with the US delegation. After a series of discussions that lasted through the night, he returned with a deal that contained little by way of real compromise from the US team. It was a re-run of previous negotiations where delaying tactics and the exhaustion of delegates were used to push watered down agreements through.

This time though, with the IPCC's Third Assessment Report confirming that human activities had contributed substantially to global warming being widely leaked and new evidence before the delegates that temperatures could rise by a catastrophic upper limit of 10.6° F (5.9° C), the EU delegation was in no mood to buy yet another compromise. Some NGO commentators are of the opinion that after the necessary numbers had been crunched Prescott's deal with the US would have led to actual emissions increases of between 6 and 9 per cent rather than any cuts. The Dutch Environment Minister Jan Pronk, who chaired the Conference, had no option but to let it close without any agreement. In a supreme irony, the venue where COP 6 took place had to be made ready immediately for an oil industry exhibition.

Reflecting the bitter disappointment of the activist community which had maintained a very visible presence both within and outside the conference center, Tony Juniper of Friends of the Earth UK declared, 'No words can truly express our anger at what has happened here, or our sadness for the victims of climate change that is to come. The world will pay the price in tears.'

Many European activists felt, despite their disappointment, that the failure of the talks was preferable to a worthless deal which they would have been hard-pressed to support. Also for once the EU had refused to submit to what they considered the US's bullying tactics. This they felt would make for a stronger position in future negotiations. There were also calls to proceed with action on climate change regardless of the

US position, isolating it as the world's greatest polluter in the global political community.

In July 2001, the use of 'sinks' as proposed by the US slunk in anyway at a deal made in Bonn (COP6.5). WWF reckoned that the 'sinks' concession effectively reduced the commitment to cut emissions by 5.2 per cent on 1990 figures to about 1.8 per cent. Further dilution came by the way of softening of compliance and penalty mechanisms.

US pulls out

With George W Bush as President, the oil industry's love affair with the Republican Party became swiftly apparent with the US pulling out of the Kyoto Protocol altogether. Bush parroted the skeptics' claim that action on emissions reduction would be damaging to the US economy. Skeptical scientist Richard S Lindzen was chosen to give the Bush cabinet a tutorial on climate change.

The rest of world has carried on negotiations without the US. At the Marrakech round in 2001, delegates met in a nation showing the ravages of disrupted weather systems to hammer out the final details of the Kyoto Protocol. Morocco had suffered a four-year drought, which had left it with depleted water tables, encroaching deserts, stalled hydroelectric resources and agricultural damage. The Kyoto Protocol had been 'saved' (albeit after much watering down) and member countries announced they would start channeling 450 million euros to developing countries for cleaner technologies from 2005. This is a tiny step when compared to the challenge ahead.

In an attempt to calm the chorus of condemnation that followed his decision to yank the US out of the Kyoto Protocol, George W Bush unveiled his very own alternative in February 2002. He laid before an incredulous world media his plan to reduce 'greenhouse gas intensity'. US industry would be urged to lower the ratio of emissions to economic output. While this could lead to increased energy efficiency there was no mention of any bottom-line reduction of emissions. California Democrat Henry Waxman was astounded at the 'doublespeak' of this plan: 'What he calls a reduction in "greenhouse gas intensity" is in reality a large increase in actual greenhouse gas emissions. And his proposed voluntary system for tracking emissions will make Enron's books look honest in comparison.' Just a year earlier Bush had announced his national energy plan which called for the construction of up to 1,900 new power plants, most to be powered by the dirtiest fuel – coal.

The Bush administration's trail of controversial environmental decisions showed no sign of petering out. In April 2002, Dr Robert Watson, chair of the IPCC, was voted out of his job in a secret ballot, with the US State Department opposing his re-election. Watson, a plain-speaking and highly regarded scientist, had fallen foul of the fossil fuel industry. Environmental groups uncovered a memo from ExxonMobil, who were major contributors to Bush's election campaign, asking for the removal of Watson whom they felt had an 'aggressive agenda'. Fortunately his replacement – Dr Rajendra Pachauri of India – turned out to be no less outspoken on the issues, even talking of 'passing a point of no return' regarding the breakdown of the world's climate. However, by 2003 ExxonMobil was handing out more than $1 million a year to right-wing organizations that opposed action on climate change.[10]

Having pulled out of the Kyoto Protocol, the US administration now proceeded to attempt to destroy it. It moved to woo various countries to sign up to bilateral climate agreements to cut emissions that observers say weren't worth the paper they were written on as there were no calls to direct action. Australia and India were among the first countries to sign on the dotted line. Indian environmentalist Sunita Narain reacted angrily:

'The US walked out of the Kyoto Protocol. It walked out of a multilateral agreement to limit luxury emissions, so that the poor would get ecological space, and the earth's climate system would recover. But the US did not merely reject the Protocol: it said it would work overtime to kill it off. The US says it will prove its strategies for "voluntary measures" – to switch to cleaner technologies – and "bilateral agreements" will be more effective than a multilateral rule-bound agreement. Forget the rules now. What the US is promising is that instead of the 5.2 per cent cut in emissions at 1990 levels, as the Protocol requires, it will increase its emissions by over 30 per cent in the agreement period.'[10]

Ratification and beyond

Ratification of the Kyoto Treaty demanded that countries accounting for at least 55 per cent of all greenhouse gas emissions stay on board. In 2004 Russia began to indicate that it might not sign up. Many thought this change of heart was prompted because Russia had been hoping to sell much of its excessive emissions quota to the US under the Protocol's carbon-trading mechanisms. With the US pullout this went up in smoke. However, eventually Russian President Putin committed Russia to the Protocol and it came into force on 16 February 2005.

The Russian dither had concentrated minds on the extent to which emissions-trading was dominating the whole Kyoto process. The future global carbon market's worth by 2010 will be gigantic – estimates range from $10 billion to $1 trillion.[11] At the 2005 Carbon Expo in Cologne, a World Bank spokesperson outlined what was happening in peachy terms: 'Companies... want to buy carbon assets in developing countries... representatives from developing countries are offering those assets.' In this global marketplace myriad schemes of dubious value get traded with little regulation and a colossal scope for fraud while the actual business of cutting emissions at source in the countries that are the worst polluters gets constantly deferred.

After ratification the Kyoto discussions rumbled on and the US, though not a Protocol signatory, still kept influencing the agenda with a little help from its OPEC friends. Discussions in Bonn in May 2005 were limited to being 'informational seminars', with the US and its allies succeeding in stopping delegates from discussing anything that might resemble an action plan.

We beg to differ

Not only does climate change affect people in the Majority World disproportionately, it affects disadvantaged people within those countries as also within the rich world even more. Indigenous people and poor communities have to put up with the excesses of the petrochemical industry, have their land and forests usurped by carbon trading schemes, and don't have the wherewithal to withstand extreme climate events. While the delegates at the Montreal meeting of the Kyoto Protocol in December 2005 were patting themselves on the back and hustling those carbon credits, there were also dissenting voices.

'We are hit first and hit hardest. In the Gwich'in community living near the oil industry in Alaska, asthma rates have rocketed to 80 per cent in the last two decades... We are also being affected by climate change as glaciers are melting. We lost 15 per cent of our caribou herds from changing weather patterns. This is important because we are one with the caribou. We have the heart of them and they have the heart of us. When one is affected it is devastation to us culturally, spiritually and physically.' – *Faith Gemmel, Indigenous Environmental Network*

'The Kyoto Protocol has put its faith in markets. How can we as indigenous peoples put our faith in these approaches when it is the market's unquenchable thirst for consuming resources that has caused the problem in the first place. ...227 indigenous groups in Alaska are affected by both the oil industry and climate change. The forests are burning, the fish are diseased, trees are dying because the permafrost they are rooted in is melting and the ecosystem is being destroyed by oil companies' activities. We don't have time to wait, the solutions must come now.' – *Clayton Muller-Thomas, Indigenous Environmental Network*

Source: Heidi Bachram, 'Kyoto fails indigenous peoples on climate justice', Carbon Trade Watch, 1 December 2005.

The meeting in Montreal in December 2005 was hailed in many quarters as a success with the industrialized countries governed by the Kyoto Protocol (39 nations, excluding the US and Australia) agreeing in principle to making deeper emissions cuts when their present targets run out in 2012. They decided to agree the new cuts by 2008. The US agreed, against its will, to 'open and non-binding' talks on new measures that all countries can pursue. Despite the vagueness of this latter agreement, it was more than many negotiators had expected from the US.

But other participants saw things differently. For some, the meeting was a thinly disguised carbon-trading jamboree. Daphne Wysham, Director of the Sustainable Energy & Economy Network, said: 'Almost everyone, from the environmental NGOs to the government representatives, and of course the overwhelming majority of business groups, is talking about the "business opportunities of climate change". All of this is conducted in the new lexicon of "carbon-trading".'

Arief Wicaksono of Ecuador-based Oilwatch International declared: 'These negotiations are all about making money of the climate crisis, not about bringing about a change in the current system of heavy subsidies for the fossil fuel industry. The truth of the global environmental crisis is beyond economic perspectives.'[12] For Fabian Pacheco, a government delegate from Costa Rica, another important issue had been ignored: 'If we are going to seriously address climate change we need to rein in petroleum companies. We need a moratorium on oil exploration because oil companies are largely responsible for the climate crisis.' Earlier that week his government had announced a moratorium on oil and mining exploration in Costa Rica.[13]

At this juncture it is worth considering what Kyoto will achieve if its first round of targets are met. According to a Dutch study, when the role of emissions-trading is accounted for, it will only succeed in reducing emissions by 0.1 per cent below 1990 levels instead of the agreed reduction of 5.2 per cent.[8] The IPCC's Third Assessment Report recommends cuts of between 50-70 per cent. In 2007, the IPCC will publish its fourth report. It is unlikely to revise its recommendations downwards. Viewed in this context the world of climate negotiations couldn't be further removed from reality.

An issue that's here to stay

It is ironic that despite the abysmal political response, the profile of the climate change issue is higher in public perception than ever before. This is in part due to the tireless efforts of environmental groups who have been vocal and creative in their protests (though many have become co-opted by market illogic along the way) and who have been key disseminators of information. It is in part due to scientists who are increasingly appearing on public and media platforms to warn of the dangers and who have often been the most eloquent voices in the argument for equitable political solutions to the problem. And it is in quite considerable part due to the weather itself as it continues to throw surprises worldwide with each passing month.

1 Ross Gelbspan, *The Heat is On* (Perseus Books, 1998). **2** 'Cry foul, cry freedom', *New Internationalist*, No April 1990. **3** Jeremy Leggett, *The Carbon War* (Penguin, 1999). **4** Christian Huot, unpublished report for the *New Internationalist*, 1999. **5** Simon Retallack, 'How US Politics is Letting the World Down', *The Ecologist*, March/April 1999. **6** Peter Bunyard, *The Breakdown of Climate* (Floris Books, 1999). **7** 'Time to close the forests and climate loopholes', World Wildlife Fund/ World Wide Fund for Nature (WWF) press release, 10 May 2000. **8** Heidi Bachram, 'Climate Fraud and Carbon Colonialism: The New Trade in Greenhouse Gases', in *Capitalism Nature Socialism*, Vol 15, No 4, December 2004. **9** 'Kyoto climate treaty off course, heading for the rocks', WWF press release, 13 June 2000. **10** Ross Gelbspan, *Boiling Point* (Basic Books, 2005). **11** Melanie Jarman, 'Emissions Trading – a desperate distraction?' *Ethical Consumer*, July/August 2005. **12** 'All Aboard the Greenhouse Gravy Train', Durban Network for Climate Justice press release, 10 December 2005. **13** 'Carbon Cartel Controls Climate Conference', Durban Network for Climate Justice press release, 10 December 2005.

7 Lasting solutions for a global crisis

How much more carbon can we send into the atmosphere... the argument for sustainability and equitable solutions... not all renewable energy is clean... a wealth of energy options... efficiency for a change.

WITH EACH PASSING year come new and more compelling discoveries about climate chaos, and the fossil fuel business becomes more profitable. The control of oil resources in particular has led to a world bristling with political instability and conflict. Meanwhile, for the oil giants it's business as usual. In 2005 ExxonMobil made the largest profit ever recorded by a US corporation – $32 billion. Shell raked in $23 billion and BP $19 billion. It's a disconnect with which we have become used to living.

It's also a disconnect about which we are not particularly happy. Knowing this, and with an eye to a very different future, Shell and BP have been investing in the renewable energy market. (ExxonMobil, it appears, couldn't give a fig.)

Some years ago *Time* magazine ran a double-page advertisement with the plaintive headline: 'Exploit or explore?' One half of the page was monochrome with a bulldozer plowing through the carnage of lush vegetation. Over it hovered the ghostly outline of the Shell logo. The other side of the page was a burst of green – trees, ferns, calla lilies sprawled across it, evocative of Eden. Over this picture, which had the kind of unreal beauty common to photographs of dishes in certain kinds of cookbooks, was stamped the Shell logo in full color.

The advertising copy did its best to suggest a caring, sharing transnational. 'Our shared climate and finite natural resources concern us as never before, and there's no room for an attitude of "It's in the middle of nowhere, so who's to know?"' it asserted. 'If we're exploring for oil and gas reserves in sensitive areas of the world, we consult widely with the different local and global interest groups. Working together, our aim is to ensure that bio-diversity in each location is preserved. We also try to encourage these groups to monitor our progress so that we can review and improve the ways in which we work.'

It was a 'pinch me I'm dreaming' moment. But around the same time there was also a TV advertisement, with a photogenic female earth scientist projecting the caring hands-on line in a tropical forest location. We were supposed to trust her, she wasn't a man in a suit; she would do right by the world.

Well, only if anyone were gullible enough to believe it. Shell has spent millions cleaning up its image and has donated money to environmental groups foolish enough to accept it. But it has a string of environmental disasters to its name and is implicated in numerous human rights violations. In Nigeria alone, Shell (along with other oil transnationals) has fed government corruption, contaminated huge swathes of land due to oil spills, and damaged the health of thousands due to its gas flaring. Several thousand people have died as a result of oil fires and pipeline ruptures, to say nothing about the violence with which resistance to the corporation has been met.

As for BP, it changed its name from British Petroleum to Beyond Petroleum; its logo is now a sunburst – or is it a green sunflower? It is trumpeting its investment in solar energy. But how far 'beyond petroleum' is it? Not much, it seems. In its year 2000 AGM, BPAmoco's own shareholders voiced concern over Northstar, its new Arctic oil field. As many as 13.5 per cent of them voted to abandon oil production in this ecologically fragile region and a redirection of those funds to BP's solar subsidiary.[1] Although by no means anywhere near a majority this still represents a significant chunk of shareholders in what is essentially a fossil fuel company. In 2005 BP's lobbyists were busy in Washington, attempting to scupper legislation within a proposed energy bill that aimed to introduce a mandatory curb on greenhouse gases in the US.[2]

Both Shell and BP are criticized by environmental groups because their investments in alternative forms of energy are not being met with corresponding cuts in their fossil fuel exploration and exploitation activities. In the light of this, their forays into solar and wind energy look like little more than cynical attempts to get in ahead of the game. They are busy patenting technology so they can capture chunks of the market and skew it to their own priorities when the rapidly expanding renewables sector really takes off.

For the time being their reality remains a polluting and polluted one. According to Professor Michael Dorsey of Dartmouth College in the US: 'BPAmoco emits more carbon dioxide than the total CO_2 emissions of the United Kingdom where the company is based. Royal Dutch-Shell emits more carbon dioxide than some of the largest countries in the world including Canada, Brazil and Mexico.'[3]

If, as is often argued, this is only the economic imperative doing its thing, then attention needs to be drawn to the larger reality of the ecological imperative. Economies are part of ecological systems and not above them, a message that business leaders don't seem to want to hear. Andrew Simms, the policy director of the New Economics Foundation, has been doing some sums. According to British Treasury estimates, the environmental damage per tonne of carbon dioxide can be quantified at around $35. Simms finds that when the environmental damage for Shell and BP's emissions and from their products is subtracted from their profits they land up respectively $8 billion and $31 billion in the red.[4] Of course, environmental health remains ultimately unquantifiable, but such calculations reveal the unsustainability of the fossil fuel economy.

No to doom

The UN Advisory Group on Greenhouse Gases estimated in the 1990s that the outer limits for temperature change to which the earth could adapt successfully were 0.2° Fahrenheit (0.1° Celsius) per decade, with

Peak oil

Once oil production has reached a maximum, it's all downhill. Experts suggest peak oil may already have been reached. Many oil companies and oil producing countries have been found to be inflating the amount of reserves they claim to possess. New discoveries of oil fields have fallen sharply. Oil exhaustion could occur around 2037 going by the trends of the last two decades.

The New Economics Foundation reports: 'In early 2004, world oil prices hovered at around $35 per barrel. An estimated $5 per barrel price rise would, at current levels of consumption, increase the energy bill of developing countries by $90 billion per year – a figure far outstripping current total overseas development assistance to poor countries. By May 2004, the oil price was steadily above $40 per barrel.' A fossil-fuel free future makes more sense than ever.

Demand outstrips discovery

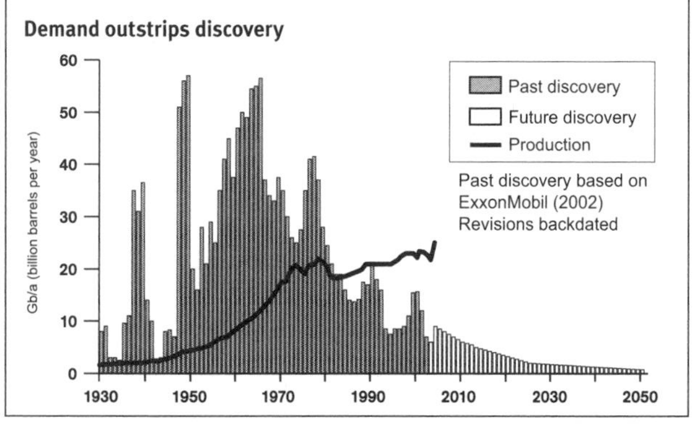

Sources: New Economics Foundation, *Mirage and oasis: Energy choices in an age of global warming*, June 2005; New Economics Foundation, *The price of power: Poverty, climate change, the coming energy crisis and the renewable revolution*, 2004.

a maximum tolerable increase of 1.8° F (1° C). Beyond this latter limit, the UN group had predicted that the earth's response to the temperature signal would be 'rapid, unpredictable and non-linear' leading to widespread ecosystem damage. The closing decades of the 20th century outstripped the decadal upper limits and we've already experienced a warming of 1.4° F (0.8° C).

Today our headlines are awash with climate crises. Now there is an urgency that we must try to control the planet's fever by not letting it exceed 3.6° F (2° C). This raised upper limit will no doubt result in further climatic ructions, but there is a slim chance we could succeed. It all depends on how quickly we manage to control our greenhouse gas emissions. Before industrialization, the atmospheric concentrations of carbon dioxide were pretty stable at 280 parts per million (ppm). Today they are 380 ppm and growing. When all greenhouse gases are taken into account, concentrations have crossed 400 ppm CO_2-equivalent already. The scientific best guess is that greenhouse gas concentrations cannot rise above 450-475 ppm CO_2-equivalent and then must fall if we are to prevent raising the temperatures above 3.6° F (2° C).

However, the rapid rate of emissions increases, the lack of political will for real change, and the predictions of some scientists such as James Lovelock (the creator of the Gaia hypothesis which postulates that the earth acts as one super organism), have persuaded many that we are already doomed. There is a fairly common belief that the damage already done to the climate system means catastrophe cannot be averted. Such thinking leads either to defeatism or a penchant for desperate solutions. Neither is of any use if we are to tackle this gigantic problem in any meaningful way.

Mainstream scientific opinion still believes that we can keep warming below 3.6° F (2° C) – if we respond quickly enough to the challenge. Even if greenhouse gas concentrations rise to 450 ppm CO_2-equivalent, much will depend upon how quickly we then reduce them again. The analogy is of an oven being warmed up – if the controls are turned to 'hot' and then quickly turned down, the oven won't reach the 'hot' stage. Climate inertia can, scientists believe, act in the same way, blunting out a peak reaction.[5]

But the big 'if' remains – will humanity react in time? Do we have the nerve to make the changes needed?

Sustainability and equity

In order to do so the climate change debate needs to bring the wider issues of sustainability and equity much more into its remit, rather than being focused on the technicalities of emissions reductions, essential as they are. Here's why.

For decades now the West has pursued the mantra of economic growth at any cost. Increases in the gross domestic product (GDP), whether they stem from the most polluting and resource-intensive industries or from the increased police activity of countering crime, have been viewed as 'growth'. This has led to a 'we consume because we can' mentality, which is straining the environmental resources of the Earth to its limits and has created chasms of inequality between nations and between the rich and poor within nations.

One US citizen laid down the consumption record of her fellow Americans in these words: 'The United States comprises 5 per cent of Earth's population and consumes 25 per cent of its resources. It is deeply painful, as an American, to state this. Yet these are the facts. Americans use twice as much water as someone from a developing country. Americans use twice as much energy as someone from Germany, France or the United Kingdom; 50 times that of a Guatemalan, 100 times that of a Vietnamese and 500 times that of a person living in Chad. With 5 per cent of the global population, we consume 25 per cent of the planet's fossil fuels each year, emitting 20 per cent of its greenhouse gases. City dwellers in the US generate twice the amount of trash as their counterparts in Spain, Italy or Germany. The US produces ten times the amount of hazardous waste as the next largest producer.

'...A typical American dinner travels over 1,000 miles from farm to table. The processing and packaging add substantial costs to the food (4 per cent of our per capita expenditure on all consumer goods goes for packaging – $225 per person a year)... In the last 200 years we have lost 50 per cent of our wetlands, 90 per cent of our old-growth forests, 99 per cent of the tall grass prairie and as many as 490 species of native plant and animal species.'[6]

The writer is no hair-shirted hippie but former US Congress member Claudine Schneider – and a Republican

Power point

A quarter of the world's population – in the North – consumes more than 70 per cent of the world's commercial energy while the remaining three-quarters – in the South – consume less than 30 per cent.

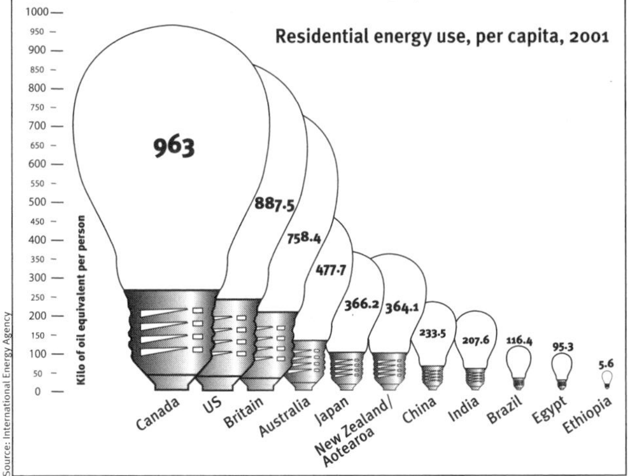

Source: International Energy Agency

to boot. According to the International Energy Agency, the average US citizen generates as much carbon dioxide in one day as a Chinese person does in more than a week and a Tanzanian in seven months.[4]

The desecration of the planet is plain to see and there can be little doubt that vastly over-consuming lifestyles cannot be sustained indefinitely. In the Netherlands, despite the great cultural value associated with thrift, it is estimated that the average citizen would have to cut their consumption by a whopping 70 to 90 per cent in order to live within their environmental space.[7] Friends of the Earth Netherlands have estimated that one-and-a-half times the country's total land area is under forest *abroad*, just to supply it with its wood and paper needs.[8] This tiny densely-populated country has a purchasing power that is two-thirds that of India, a land 93 times its size. Viewed in this light, the climate crisis is but one more manifestation of unsustainable growth, albeit one which could have devastating effects in the near future. If we are to build a future, where the threat to the environment could recede, sustainability must be a key consideration.

No level playing field

The equity argument follows on from the sustainability argument. There cannot be a level playing field in emissions reductions when there are such gross inequities in the wealth of nations. Much of this wealth has been derived from environmental degradation in the first place. Andrew Simms proposes that the industrialized world is in fact greatly indebted when environmental issues are brought to bear on conventional economics. 'The logic of a radical change of behavior, in the face of what is a huge environmental debt to the global community, should not be difficult to grasp in the industrialized world. The rich countries have been demanding for decades that very poor countries undergo massive change in response to much smaller financial debts... There is a direct link between fossil fuel use and economic output. Because of this, the carbon debt can be given illustrative values in economic efficiency terms. Such sums show the heavily indebted poor countries actually in carbon credit up to three times the value of their ordinary debts.' Simms estimates that the seven most industrialized nations of the world have an ecological debt totaling $13 trillion.[9]

An essential element of climate negotiations should be the recognition of this environmental indebtedness. This would allow for stricter emissions targets for industrialized countries and the massive transfer of clean technologies to developing countries, which would aid them in switching to sustainable, renewable energy in their quest for economic security. Such measures would require thinking that went beyond short-term national interests to hammer out a schedule of global governance that would ensure an equitable, sustainable, energy future.

In the normal run of things there is about as much chance of this happening as there is of pigs flying, but the alternative of global bullying does not work in climate negotiations. A rising tide of activist opinion is on the side of such solutions because they realize this is the true *realpolitik*, not the dodging and diving practiced by their politicians.

False alternatives

With the climate debate on fire, the alternative energy industry sees a rosy future ahead. A ring of cash registers is heard as the burgeoning markets of India, China and Eastern Europe are contemplated. The industry chiefs have visions of power, of becoming transnationals that could market technologies that are not tied to an extractable resource but to free wind and sunlight. While the energy revolution could provide the alternative energy sector with a huge growth spurt, viewing developing countries purely as markets would wreck chances of emissions reductions. The means to develop alternative energy sources need to be decoupled from ability to pay.

The natural gas and nuclear industries are vying to paint themselves as sources of clean energy. While natural gas, with its lower carbon dioxide emissions, could have a small role to play as a transitional fuel, nuclear energy cannot seriously be considered as an option. It has certainly had a boost from some doom-ridden environmentalists who are willing to risk its dangers in order to avert climate meltdown. But here they are mistaken. First off, short of building a nuclear power station that had zero scope for error, which could remain in operation indefinitely, and was indestructible were a natural calamity like an earthquake to occur, there is no way the highly radioactive plutonium-rich fuel used could be considered safe.

The waste product from such a plant would also need to be effectively sealed off for hundreds of years before it no longer posed a threat to human health. Shutting down existing plants can take half a century and cost billions. Even with all high-level radioactive waste removed, such plants would still require monitoring for a further 300 years. Nuclear energy is primarily used for power generation. In order to switch to nuclear the world would need to build 3,000 nuclear reactors by 2100. But it won't come to that for a number of reasons. Surely a most compelling one is that studies show that if all fossil fuel generated power were to be switched to nuclear the world would have run out of economically viable uranium within just three to four years.[10]

Also moving out of the alternatives picture are hydroelectric projects involving reservoirs and dams. The dam builders have pushed hard for hydroelectricity to be included in the Kyoto Protocol's list of sustainable technologies. Long under attack from environmentalists and local communities for destroying homes and habitats, such projects also cannot claim to produce greener energy. The bad news is that reservoirs are leaking methane and carbon dioxide into the atmosphere, especially in the tropics where the submerged vegetation rots more rapidly.

Researchers from Canada, where some of the world's largest hydroelectric projects are located, claim that these are responsible for up to a fifth of all methane emissions and their total contribution to the anthropogenic [human-activity-induced] greenhouse effect could be as high as 7 per cent. At present it is believed that the third of the world's reservoirs that are located in tropical regions contribute 80 per cent of emissions from big dam projects, but in the long run reservoirs in Northern countries could begin to catch up. Reservoirs that have been built over peat bogs cause special concern as a thick peat bog could hold greater reserves of decomposable carbon than a rainforest.[11]

Another non-starter is the option being clasped most tightly to the industrialized world's bosom – carbon trading (see chapter 6). A segment from the Durban Declaration on Carbon Trading signed in 2004 by a coalition of environmental and indigenous peoples' groups gets to the essence of what's wrong with it: 'History has seen attempts to commodify land, food, labor, forests, water, genes and ideas. Carbon trading follows in the footsteps of this history and turns the earth's carbon-cycling capacity into property to be bought or sold in a global market. Through this process of creating a new commodity – carbon – the earth's ability and capacity to support a climate conducive to life and human societies is now passing into the same corporate hands that are destroying the climate.'

Each year the sun shines down the energy equivalent of 1,000 trillion barrels of oil.[12] Thomas Edison (1847-1931) had tremendous faith in it: 'I'd put my money on

The carbon capture dream

With emissions stubbornly refusing to go down at source, the feasibility of storing carbon dioxide safely has exercised scientific minds.

An international team led by Philip Boyd of New Zealand's University of Otago succeeded in creating a phytoplankton bloom over 90 miles (150 kilometers) long and 2 miles (4 kilometers) wide in the Southern Ocean by scattering over 18,000 pounds (8,500 kilograms) of iron salts in the water. They were acting on an idea mooted 10 years earlier of boosting the sea's carbon dioxide absorption by promoting the growth of phytoplankton which draw the gas down from the atmosphere.

This artificially created bloom lasted for a month, but left little evidence that the carbon dioxide drawn down by phytoplankton had actually sunk to deeper waters, something that would need to happen if the aim is to create a permanent sink of the gas.

The possibility of sinking more carbon dioxide in the oceans either by diffusion in the waters or by forming CO_2 lakes on the deep ocean bed has also been explored. Leaving aside the enormous financial constraints, both options would lead to further acidification of the oceans, spelling disaster for all the creatures that live in them – to say nothing of the gas leaking back out.

Scientists are more confident that carbon dioxide could be better contained underground at depths below 800 meters (2,625 feet) if there were physical and geochemical trapping mechanisms. A caprock [dense, impermeable rock layer capping a mass of porous material] would be essential. Costs would be high though, both of capturing the carbon dioxide from industries in forms that could be storable and of injecting it deep below the earth's surface. It would be far wiser to invest the money into reducing emissions. In the event of a sudden leakage (say due to an earthquake) all the effort would be wasted and living beings in the vicinity would die if the concentrations were high enough. ■

Sources: Steve Connor, *The Independent*, 13 October, 2000; IPCC Special Report, *Carbon Dioxide Capture and Storage*, 2005

the sun and solar energy. What a source of power! I hope we don't wait 'til oil and coal run out before we tackle that.' The solar potential for energy production is at least 1,000 times more than the energy currently being used worldwide. Photovoltaic (PV) cells and panels which produce electricity directly from sunlight can enable a building to generate more electricity than it needs – the technology already exists. In Britain it is estimated that a typical house fitted with solar panels would save up to 2.5 tonnes of carbon dioxide emissions a year. Up until recently the silicon-based cells were relatively expensive to manufacture but Japanese technology has paved the way for super-thin cells that considerably save on raw materials. Costs are now down to less than a tenth of what they were 20 years ago.

The Japanese Government encouraged household solar panels in the 1990s by offering subsidies. Today Japan is the number one producer of solar power among all nations. Subsidies are now being phased out, but capacity is still expected to grow at an extra 20 per cent a year. Similar joined-up government thinking is to be found in Germany, where energy companies are legally bound since 1991 to buy any renewable power anyone generates at a generous price. Since then the country's solar capacity has been growing by nearly 50 per cent every year and 10,000 people are employed in the sector.[13] The German solar panel market is diverse – around 300 companies supply it, preventing market domination by any one. The US has plans to install a million solar roofs by 2010 and growth of PV generation is expected to be nearly 20 per cent a year for the next two decades.

Where funds have been available, Majority World countries have been quick to harness solar energy. In Mongolia some of the *gurs* (tents) of nomadic herding communities have solar panels parked outside them. Where there is sunlight there is solar energy to be had, and it doesn't have to be warm for them to work – in fact they are most efficient in colder temperatures and snow reflects additional light onto them. A single 40/60-watt solar module can provide a nomadic family with five hours light, and energy for a television and radio. In Eritrea, the Democratic Republic of Congo, Sudan, Uganda and other African countries solar power is being harnessed in areas remote from the electricity grid to keep medicines and vaccines cool in clinics.[14] In Kenya more rural households get their electricity from solar energy than from the national grid.

Large-scale solar generation has remained a pipe-dream. However, with critics claiming such energy would be more expensive. But with no major effort made so far to build a large enough plant that would generate enough power to lower unit costs, this argument rings hollow. As solar champion Jeremy Leggett put it: 'How much would it cost to build a price-busting plant? The answer is about $100 million. A single oil rig can set you back $4,000 million these days. We are talking about less than one leg of an oil rig to show the world that electricity can be generated by the sun as cheaply as it can by burning fossil fuel, pretty much anywhere we want it, even in cloudy latitudes.'[15]

Other systems of solar technology use the sun's energy to directly heat water – all new buildings in Israel are required to have such installations, cutting the country's fuel imports by 5 per cent.

The sky's the limit

Renewables have the potential to fulfill the world's energy needs several times over.

Usable global renewable resource base (Exajoules* per year)

Resource	*Current use*	*Technical potential*	*Theoretical potential*
Hydropower	10	50	150
Biomass energy	50	> 250	2,900
Solar energy	0.2	> 1,600	3,900,000
Wind energy	0.2	600	6,000
Geothermal	2	5,000	140,000,000
Ocean energy	–	–	7,400
Total	**62.4**	**> 7,500**	**> 143,000,00**

* 1 exajoule = 1 million, million megajoules

Source: New Economics Foundation, *The price of power: Poverty, climate change, the coming energy crisis and the renewable revolution*, 2004.

The wind and the waves

Solar is by no means the only clean energy source. Windmills have been used to pump water and grind grain in Holland for hundreds of years. Today though, Dutch farmers are encouraged to invest in wind turbines that catch the wind's energy with their propeller-like blades and turn it into electricity by mechanically spinning a generator. Any excess electricity can be sold back to the national grid.

Wind power is economic as well as ecological, because once the initial costs of the turbine are covered, running costs are quite low. It can also reach areas where the outlay for conventional power lines connected to a central grid would be prohibitive. A southern Indian wind farm with 2,000 turbines is the second largest in the world. China has the world's largest wind power capacity at an estimated 253 million kilowatts. The country has made huge strides – an earlier projection had suggested that only eight million kilowatts would be generated by wind power in the country by 2020. Now it reckoned that by that date the country will provide a full 10 per cent of its exploding energy needs by wind power. It has been estimated that Britain has the largest offshore wind resource in Europe, capable of generating three times the nation's electricity requirements. Greenpeace UK calculates that if Britain made the commitment over the coming decade to generating 10 per cent of its electricity from this source it could lead to the creation of 30,000 new jobs.

Whereas large-scale dams and reservoirs cannot figure in the clean energy equation, micro-hydro projects that channel smaller bodies of water through turbines could still go some distance towards solving the energy needs of smaller communities. The oceans could be tapped both for thermal energy and the mechanical energy of tides, which could be converted into electricity.

The worldwide potential for wave power is gigantic, with ideal locations for tidal power stations along the western coasts of Europe and the US, the coasts of New

Zealand and Japan. Energy could be generated by placing bell-like structures in the sea, which when the waves entered or left them would expel or pull in air through a turbine located at the top the structure. The movement of the turbine would generate electricity. Tidal barrages at the sea mouth or in rivers could gather water at high tide allowing it to flow back through a power-generating turbine when the tide moved out or river levels fell.

There are a number of replacement fuels waiting in the wings to take over from oil and coal that could help cut emissions. Biogas, which comes from farmyard waste and burns relatively cleanly is being used by Indian and Vietnamese farmers. But such fuels must be viewed as transitional rather than fully-fledged renewables. It is argued that biofuels are cleaner than coal or oil because the plants they are derived from absorbed carbon while they grew; coal and oil liberate historical stores of carbon. But biofuels release emissions nonetheless and their promotion is not straightforward for other environmental reasons.

In Brazil's case, President Lula is keen to promote renewable energy. But one of the options on the table is an environmental nightmare – biodiesel derived from soya beans. Besides the fuel not having an edge over conventional fuels in terms of emissions, this 'solution' would create massive deforestation to clear land for soya plantations.[16]

Hydrogen derived from water is the cleanest fuel of all and could power homes as well as cars. Hydrogen gas combines with oxygen in a fuel cell to produce energy. Its waste products are heat that could be used for heating purposes and water. Most major car companies are researching fuel cell vehicles. The electricity required for isolating the hydrogen could be derived from renewable sources. Sadly this is not the plan of the big oil companies who are gearing up to produce it from natural gas.

With such a multitude of energy options available – and many of the technologies have been around for decades crying out for further development – it seems like nothing short of a death wish that governments across the globe spend over $235 billion each year subsidizing the fossil fuel industry.[17] If such subsidies could be diverted to clean, renewable energies along with revenues from environmental taxes, the transformation in the energy scenario could be revolutionary.

Efficiency

But while we're holding our breaths for the revolution, a bit of common sense may not go amiss. The field of energy efficiency has massive emissions reductions and financial savings to offer. Former US President Clinton went on record as saying that a 20-per-cent reduction in emissions could be achieved in the US just by wasting less energy.[18] Other studies have put the percentage even higher than that. In Britain appliances left on the standby mode create a million tons of carbon emissions – enough to keep the country's street lights burning for four years.[19] In 2001 Brazilians were threatened with power cuts. They voluntarily dropped consumption by 20 per cent without any perceptible reduction in quality of life. In many cases less really can be more. Many people of my parents' generation who lived through World War II see thrift as the key to enjoying a full life. Where avid consumerism isolates, sharing resources builds community.

Efficient home building has often been a key feature of temperature control in many poor warm countries where there may either be no electricity or it may be too expensive to use for cooling purposes. Courtyards at the center of the house or cooling towers above them help to lower temperatures without expending any energy. Trees throwing shade onto walls keep them cooler. Research on similar smart buildings in the Western world suggests that architects need to pay more attention to simple things like access to sunlight in order to save electricity costs in buildings. Commonsense ideas such as building offices with an optimum depth so that every desk has some access to daylight can lower lighting costs dramatically. The clever design of appliances to use minimum amounts of energy can save two-thirds of the electricity used.

Efficiency can also be extended in the sphere of public transport, such as the much-lauded integrated bus system in Curitiba, Brazil, which has lowered car use, reducing petrol consumption by 30 per cent. 70 per cent of Curitiba's citizens use buses.

The remarkable thing about efficiency is that a little goes a long way, so people can actually live better for less. At a personal level efficiency need not involve rocket science – it can be as simple as remembering to switch off all appliances that aren't being used and reusing and recycling more.

But consumption is the key – whether at the personal level or at the level of commerce or government. Unless there is a willingness to consume less and more wisely, there's little chance of stopping the tornado of climate change. This doesn't mean hair-shirt environmentalism or a return to the Stone Age (indeed technical advances will play a major part in any energy transition); it will mean radical lifestyle changes and a commitment to environmental equity.

Currently emissions from transport are proving to be the most intractable area where deep energy cuts could be made. Air travel, continually growing, threatens to wipe out progress in energy efficiency elsewhere. Developing hydrogen fuel cells for airplanes doesn't appear to be an answer – they would release water vapor in the atmosphere, thus contributing to global warming. Until a cleaner technology arrives, we must fly less.

However, another logic is increasingly prevailing among the environmentally conscious. This is the thought that they can buy their way out of their pollution. Companies providing carbon offsets are thriving. Take that long haul flight anyway – then just pay someone to plant trees in the Majority World. Apart from the

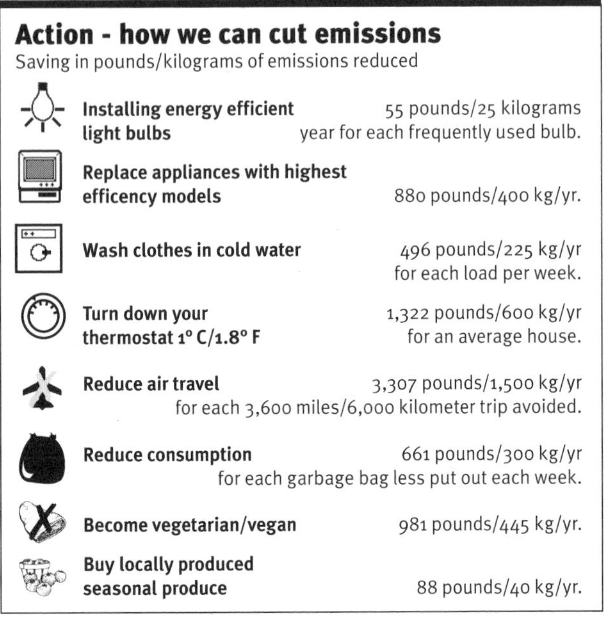

Source: Canadian Housing Information Center

dubious projects associated with much carbon offsetting, this way of thinking betrays a lack of understanding the issues at stake. Soumitra Ghosh, from the National Forum for Forest Peoples and Forest Workers in India, who are fighting offset projects in the country, sets the record straight: 'We have nothing against planting trees or people helping communities in "poor" countries. But if such actions mean that "actual" and measurable emissions of greenhouse gases would continue as usual under the safe cloak of "offsets" and some people would earn dollars and euros out of such supposedly "green" and "environmental" actions, perhaps it's time to tear the cloak, once and for all." Climate activist Adam Ma'anit offers a succinct summation of what's really needed: 'Fly less, buy less, regulate polluters and support communities affected by pollution and climate change. The real solution to climate change is social change.'

There are answers enough from the tiniest to the grandest scale, ranging from the most basic technology to the most sophisticated, but they can never work within a global political framework that pays lip-service to sustainability and which has at its foundations colossal inequity. Left to their own devices there seems to be little evidence that the politicians of some of the world's richest nations will do anything decisive. Here too, we have a mandate as individual citizens to call them to order. It will need a groundswell of ordinary citizens making their objections to the status quo plain in order to 'wrestle the earth from fools' (as singer Patti Smith put it). The nascent climate justice movement has begun to voice the equity arguments with clarity. You could add your voice to theirs.

1 'Shareholders vote against BP Arctic oil plans', *Positive News*, Summer 2000. **2** Katherine Griffiths, 'BP shows two faces as it fights US bill to cut CO_2 emissions', *The Independent on Sunday*, 12 June 2005. **3** 'Carbon Cartel Controls Climate Conference', Durban Network for Climate Justice press release, 10 December 2005. **4** Andrew Simms, 'Carbon addicts and climate debt', 9 February 2006, http://news.bbc.co.uk/1/hi/sci/tech/4696924.stm **5** Malte Meinshausen, Reto Knutti and Dave Frame, 'Can 2° C warming be avoided?', 31 January 2006, www.realclimate.org **6** Claudine Schneider, 'Consumption: United States', *The Planetary Interest* edited by Kennedy Graham, (UCL Press 1999). **7** Resource Renewal Institute, 'The Netherlands' National Environmental Policy Plan', 1998. **8** Friends of the Earth Netherlands, *Sustainable Netherlands Revisited*, 1996. **9** Andrew Simms, 'A climate of debt', *Resurgence*, July/August 2000. **10** Adam Ma'anit, 'Nuclear is the new black', *New Internationalist*, September 2005. **11** Fred Pearce, 'Plumbing the depths of insanity', *The Independent*, 13 October, 2000. **12** Dave Mussell, Juleta Severson-Baker and Tracey Diggins, *Climate change: Awareness and Action* (Pembina Institute for Appropriate Development, 1999). **13** Adam Ma'anit, 'Renew Yourself', *New Internationalist*, September 2005. **14** Vanessa Baird, 'Leapfrogging', *New Internationalist*, October 1996. **15** Jeremy Leggett, 'Solar PV: Talisman for hope in the greenhouse', *The Ecologist*, March/April 1999. **16** Giulio Volpi, 'Soya is not the solution to climate change', *The Guardian*, 16 March 2006. **17** New Economics Foundation, *The price of power: Poverty, climate change, the coming energy crisis and the renewable revolution*, 2004. **18** United Nations Associations of the USA website http://www.unusa.org/programs/gcc.htm **19** Ben Russell, 'Standby Britain: How it fuels our energy crisis', *The Independent*, 23 June 2005.

Contacts

INTERNATIONAL

Carbon Trade Watch
Research group.
www.carbontradewatch.org

Climate Action Network
Co-ordinates climate NGO activities.
www.climatenetwork.org

Climate Ark
Vast archive.
www.climateark.org

Climate Indymedia
www.climateimc.org

Durban Network for Climate Justice
International network campaigning for climate justice.
www.carbontradewatch.org/durban/durbandec.html

Friends of the Earth International
Lobbying for action on climate change.
www.foei.org

Greenpeace International
Pressuring governments and industry to shift from dirty energy.
www.greenpeace.org/~climate

The Corner House
Research on climate policy.
www.thecornerhouse.org.uk/subject/climate/

The Intergovernmental Panel on Climate Change (IPCC)
www.ipcc.ch

Oilwatch International
A Southern resistance network.
www.oilwatch.org

Planet Ark
Environmental daily news portal.
www.planetark.org

Real Climate
Climate science weblog.
www.realclimate.org

Sinkswatch
Tracks carbon sink projects.
www.sinkswatch.org

World Rainforest Movement
Network to defend the world's rainforests.
www.wrm.org.uy

AOTEAROA/NEW ZEALAND

Friends of the Earth New Zealand
PO Box 5599, Wellesley Street, Auckland
tel: +64 9 3034319
fax: +64 9 3034319
email: foenz@kcbbs.gen.nz

Greenpeace Aotearoa/New Zealand
www.greenpeace.org.nz

Rising Tide Aotearoa/New Zealand
www.risingtide.org.nz

AUSTRALIA

Friends of the Earth Australia
www.foe.org.au

Greenpeace Australia/Pacific
www.greenpeace.org.au

Rising Tide Australia
http://risingtide.org.au

CANADA

David Suzuki Foundation
www.davidsuzuki.org

Energy Action
North American youth and student network.
www.energyaction.net

Greenpeace Canada
www.greenpeacecanada.org

The Sierra Club
www.sierraclub.ca

UK

Friends of the Earth England, Wales and Northern Ireland
www.foe.co.uk

Friends of the Earth Scotland
www.foe-scotland.org.uk

Hadley Center
Government research center.
www.met-office.gov.uk/research/hadleycentre/index.html

Rising Tide
http://risingtide.org.uk

Stop Climate Chaos
www.stopclimatechaos.org

USA

Indigenous Environment Network
www.ienearth.org

Just Transition Alliance
www.jtalliance.org

Sierra Club
www.sierraclub.org

Worldwatch Institute
www.worldwatch.org

Index

Page numbers in **bold** refer to main subjects of boxed text and maps

Index

Index